T0321602

WHERE DID WE COME FROM?

Life of an Astrobiologist

WHERE DID WE COME FROM?
Life of an Astrobiologist

Chandra Wickramasinghe

Buckingham Center for Astrobiology, University of Buckingham, UK

&

University of Peradeniya, Sri Lanka

Edited by

Kamala Wickramasinghe

World Scientific

NEW JERSEY · LONDON · SINGAPORE · BEIJING · SHANGHAI · HONG KONG · TAIPEI · CHENNAI

Published by

World Scientific Publishing Co. Pte. Ltd.

5 Toh Tuck Link, Singapore 596224

USA office: 27 Warren Street, Suite 401-402, Hackensack, NJ 07601

UK office: 57 Shelton Street, Covent Garden, London WC2H 9HE

British Library Cataloguing-in-Publication Data
A catalogue record for this book is available from the British Library.

WHERE DID WE COME FROM?
Life of an Astrobiologist

ISBN 978-981-4641-39-5
ISBN 978-981-4641-40-1 (pbk)

In-house Editor: Christopher Teo

Typeset by Stallion Press
Email: enquiries@stallionpress.com

Printed in Singapore

For Clement and Reuben

PREFACE

From time immemorial man has been in pursuit of answers to questions relating to his origins, his place in the Universe and to the nature of the Universe itself. At the deepest level of inquiry, the imagination of the poet, the artist, the philosopher, the composer and the scientist coincide — they all seek answers to these same questions using their different tools.

This quest for discovery can sometimes turn back on itself and rather than unearth any truths, it merely reveals its own process, and here ultimate reality poses to be ever elusive. At such times, science and mathematics with their pretences to objectivity can seem less able than music, poetry or art to explain our existence — the arts being able to offer an experience of the mystery without a self-conscious focus on absolute truths.

As humans it seems as if we are, and always have been, hard-wired to search for meaning. In the cave paintings from Palaeolithic times, a need not only to depict reality but also to imbue it with significance and meaning is strikingly apparent. In the modernist era, the 1897 Gauguin painting bares this out with the title *D'où Venons Nous/Que Sommes Nous/Où Allons Nous* — (Where Do We Come From? What Are We? Where Are We Going?)

In my long career as an astronomer I have sought to answer precisely the same questions using the craft of mathematics. The answer to Gauguin's first question is "from space — from the depths of the Universe". The answer to the second question is "an assembly of cosmic genes — cosmic viruses — no more, no less". The answer to third question is "inevitably back into the cosmos".

In the following chapters I shall describe how I stumbled upon the path that led to these answers, and how the terrain was often rough and treacherous. It has now become clear that space rocks (meteorites) and space dust hurtling across the universe could transfer life over a vast cosmic scale. The American poet Edna St. Vincent Millay saw this coming decades ahead of any scientific proof:

> "Upon this age, that never speaks its mind
> This furtive age, this age endowed with power.....
> Upon this gifted age, in its dark hour,
> Rains from the sky a meteoric shower
> Of fact...they lie unquestioned, uncombined,
> Wisdom enough to leach us of our ill
> Is daily spun; but there exists no loom
> To weave it into fabric."
>
> Edna St. Vincent Millay from *Collected Sonnets*
> (NY: Harper Perennial, 1988 p140)

CONTENTS

Author's Foreword

The idea of this autobiography was conceived at the time I was writing my book "A Journey with Fred Hoyle" (World Scientific, 2010). Because my own scientific career is inextricably linked with that of Fred Hoyle, the reader will notice several points of overlap with "A Journey with Fred Hoyle". This does not, however, detract from the separate identity of this book "Where Did We Come From?" as the story of my life and my own personal quest for understanding our origins.

Chandra Wickramasinghe
Cardiff, 2014

Maternal grandfather, Benjamin Soysa

Dionysius Lionel and Agnes Beatrice Wickramasinghe

Paternal grandparents, Dionesius Lionel and Agnes Beatrice Wickramasinghe

Parents, Percival Herbert and Theresa Elizabeth Wickramasinghe

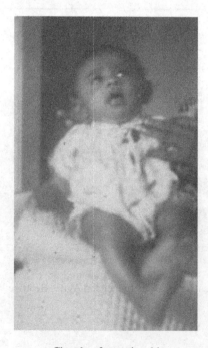

Chandra, 2 months old

Chandra, 4 years old

Dayal, Chandra, Sunitha on board ss Synthia, 1946

My home-made telescope to observe 1955 total solar eclipse, and image of partial eclipse

Galle Road, Bambalapitiya in 1956

Paternal grandparents house, 4 Francis Road, Wellawatte, 1956

35 Hildon Place, Bambalapitiya in 2000

Outside the Institute of Fundamental Studies in 2014

Royal College Colombo, 2014

Royal College Prize Day, 1956 — receiving award from Governor General Sir Oliver Goonetilake

Royal College Hall in 2014 showing prize panel bearing name N.C. Wickramasinghe

On board ss Orcades, 1960 on the way to Cambridge

In Cambridge, 1961

Trinity College, Cambridge

Nuclear physicist Houtermann and Fred Hoyle at Varenna. 1961

Punting on the river Cam, 1961

Walking with Fred Hoyle in the Lake District, 1961

On the walk with Fred Hoyle in the Lake District, 1961

My father visiting Cambridge in 1962

Jesus College Chimney entrance

With brother Sunitha at a feast at Jesus College (1963)

Priya in Ceylon, 1965

Chandra and Priya on board ss Oriana, 1966

With Priya at a Jesus feast, 1973

Hoyle family at Gregynog, 1977, flanked by Kamala and Anil

In discussion with Fred Hoyle in Cardiff, 1980

Fred Hoyle delivering a lecture at Cardiff, 1980

University College Cardiff Principal Bill Bevan, Fred Hoyle and Chandra, at the opening of the University College Cardiff Press in 1981

Zdnek Kopal at the IFS conference in Sri Lanka, 1981

With Fred Hoyle at the IFS conference in Sri Lanka, 1981

Presenting "Evolution from Space" to President J.R. Jayawardene, 1983

Presented with Honorary Citizenship of Arksansas, by the State Governor, 1984. Infant in Priya's arms is daughter Janaki

Fred Hoyle and Chip Arp at conference on Cosmology in Cardiff, 1989

President R. Premadasa decorating Chandra with the Sri Lankan National honour Vidya Jyothi, 1992

Shiela Solomon's bronze statue of Fred Hoyle, unvieiled at the Institute of Astronomy in Cambridge in 1992

President J.R. Jayawardene and Mrs Jayawardene with Priya, 1992

My photograph of Comet Hale-Bopp taken outside our house in 1995

President Daisaku Ikeda decorating Chandra with Soka University's DSc degree in 1996

In deep conversation with Sir Arthur C. Clarke in Colombo, 1996

My last photograph and memory of Fred Hoyle, in Bournemouth, 2001

Phil Solomon and Chip Arp at Fred Hoyle memorial conference in Cardiff, 2002

Presenting a copy of "A Journey with Fred Hoyle" to President Mahinda Rajapakse, President of Sri Lanka, in 2005

Priya, grandsons Reuben and Clement, and daugher Kamala in Sri Lanka, 2014

Clement, Chandra, Reuben, 2014

With Milton Wainwright in Graz, Switzerland, 2014

PROLOGUE

I started my life in a conservative culture in Ceylon (now Sri Lanka) where to conform was the norm and to rebel a heinous sin. My career in science, however, took an opposite turn at an early stage, provoking rebellion even to the extent of ultimately acquiring the status of a maverick. I did not choose such a calling but this is the way my scientific career and research naturally led me. My mentor, friend and long-term collaborator Fred Hoyle always reminded me that if solutions to important problems in science were to be found within the realms of orthodoxy, they would already have been discovered. For the biggest unsolved problems, for example the origin of life or the origin of the universe, it would not be surprising to find solutions that lie in the realm of the heterodox.

Throughout history the most important scientific and intellectual advances have come from outside the realm of conventional belief. The greatest innovators who challenged established beliefs were invariably execrated for doing so. In ancient Greece, the philosopher Anaxoragas maintained that the sun was a red hot stone, and that the moon was made of Earth, so challenging the generally accepted divine status of these heavenly bodies. For this impiety he was banished from Athens. Others who dared question hallowed beliefs have suffered worse fates: some were even burnt to death. Even in modern times challenging orthodox ideas in science is a risky business that carries penalties, although the penalties are not as extreme as they were in the past.

My life's work in astronomy first sought to discover what cosmic dust is made of. And from such a seemingly innocent project I went on to argue that cosmic dust was connected with life itself. Life did not start here on Earth, as conventional science stipulated, but was derived from a vast cosmic

1

system. Our planet was simply a building site, one of billions or trillions, for the assembly of complex genetic structures that were introduced from the external universe. I have vigorously defended this point of view because I firmly believe it to be true. Throughout my career I have seen these ideas being vindicated, bit by bit, to the point when a huge paradigm shift has become inevitable.

In this book I trace the important events of my life that led up to this position, the people who influenced me along the way, and the circumstances that enabled my progress through a lifetime.

CHAPTER 1

TO BEGIN AT THE BEGINNING

My native country, Ceylon, as I first knew it, now Sri Lanka, was known earlier by various names — Serendip, Taporbane — amongst them. The island is believed to have been inhabited by various tribes from about 30,000 years ago, but my own ancestors, the Sinhalese, were said to be of more recent origin, having migrated from North India at around 500BC, more or less contemporaneously with the classical period of ancient Greece. Magnificent edifices, ruined cities and palaces that adorn the island dating back to the 3rd century BC testify to a great civilization that existed at this time and a prosperous people trading with Greece, Persia, Arabia and China.

A succession of dynasties of Sinhalese kings ruled over the island which enjoyed many centuries of self-sufficiency and prosperity before the later foreign conquests by the Portuguese, Dutch and finally the British. The Venetian explorer Marco Polo, who visited Ceylon in the 13th century, extolled its natural beauty describing it as a "jade pendent in the Indian Ocean". Today Sri Lanka is considered as one of the most desirable tourist destinations in the world. Colombo, the present commercial capital, is a bustling, overcrowded, if somewhat polluted modern metropolis — serving as a gateway to ancient ruined cities, as well as to wild life sanctuaries and beach resorts that attract many thousands of visitors each year.

If one can be transported to a time in the late 1930's and early 1940's, when I was a child, the scene would have looked dramatically different from what it is now. Bullock carts, rickshaws pulled by men, bicycles, a few old Austin and Morris automobiles, and the odd bus or two would have vied for their place along the main roads of Colombo. There would be evidence of a highly stratified society, people revealing their social status by their dress as well as by their mode of transport. Middle class folk would dress immaculately and ride in chauffeur driven cars. The poor would be generally ill-clad and walk under umbrellas in the hot sun, or stay in long lines for the few private busses that plied for hire.

I have a vivid memory of the house where I was born and spent my early years. No. 35 Hildon Place, Bambalapitiya in a quiet suburb of Colombo was a colonial-style bungalow built on a spacious plot of land near the end of a wooded cul-de-sac that branched off from what is now the main road between Colombo and Galle — Galle Road. A short walk down Hildon Place, a gravel road full of ruts, and across the tarmaced Galle Road, led to the sea — the great Indian Ocean.

Although our house was about half a mile from the coastline, I could hear the sound of sea filter like a whisper through the quiet air of the night. With the low volume of the traffic there was hardly any air pollution as we now have, and very little in the way of light pollution. So the night sky that I could see from our front garden was incredibly spectacular — the Milky Way with its billions of stars gracefully arching across the vault of the heavens. It was such early memories that most probably beckoned me to a career in astronomy in later life.

The environs of our house were also delightfully rural and peaceful in those times. Adjoining the back garden was a thicket — a mini forest — that no one dared explore. There were often sounds of rustling of leaves heard in the stillness of the night that gave fodder to our vivid childhood imagination. But the only wild life we ever saw was the occasional small iguana (thalagoya) and a harmless rattle snake or two. What affected us more from the adjoining wild area was the existence of stagnant rainwater puddles that served as breeding ground for mosquitos. I remember we had to sleep under mosquito nets. But in spite of this precaution I believe that we all must have succumbed to a mild form of malaria — a mosquito-borne parasitic disease that still affects a few parts of Sri Lanka and swathes of Asia and Africa.

The causative agent of malaria is a protozoan (a single celled protozoa) that is transmitted by the female mosquito of the Anopheles mosquito. The strategy of preventing outbreaks of the disease is to remove breeding ground of the mosquito which is stagnant water. The treatment for malaria in those days was confined to the use of the bitter drug quinine, which occurs naturally in the bark of the cinchona tree. I remember being given this bitter powder. Nowadays, more powerful drugs are available to treat malaria, and also antimalarial prophylactics. But for myself I think that early exposure has conferred a measure of immunity that still protects me from further attacks.

To add to the wilderness that adjoined our garden, I recall a time in my early childhood when we kept poultry for eggs, and at a later stage we also kept a goat for milk. I was told that all this was done in an attempt to combat post-war austerities that were in force in the late 1940's. In the front garden we had a mango tree and several papaya and banana trees all of which flourished with little or no attention. In addition we had a pond with gold fish in the centre of our garden, a rose bower and rose trees which my father imported directly from England. He prided himself as a rose grower, and a member of the UK National Rose Society.

The peace and tranquility that I associated with my home in those days would be hard to imagine in the 21st century. The stillness and quiet of a warm sultry afternoon was interrupted only with the occasional incantations of a vendor who would go past our front gate announcing their ware with shouts of "malu, malu, malu" for example — the man who was selling fish. The fish was carried in a pair of baskets hung from the ends of a long pole that was skillfully balanced at its fulcrum on the shoulder. For such people life was hard and rough and such sights as these highlighted the social inequalities that persisted at the time.

There was no TV in those days, so it would have been our habit to spend long evenings after supper sitting on rattan chairs or lazing on arm chairs on the veranda, reading, talking, or just reflecting on the day's events in quiet contemplation. We would hear the chirping sound of the crickets and see the faintly flickering lights of fireflies merge with the stars twinkling in the night sky.

On the day I was born, 20th January 1939, Adolph Hitler proclaimed his intention to exterminate all Jews in Europe. My native country at this time was an inconsequential British colony — an appendage of the "Jewel of the Crown" which was India. This great and prosperous Empire, upon which it was said that the "Sun would never set", was beginning to show signs of strain in its social and administrative structure and appeared to be heading slowly towards its ultimate collapse. Although the wireless and telegraph were already invented, communication between the different parts of the empire was painfully slow. The fastest mode of travel in those days was by steamship, and it took several weeks to sail from London to India or Ceylon. It was this slowness of communication and contact that severely weakened the cohesion of a vast and scattered empire. Furthermore the rulers of the Empire and the holders of power were not yet accustomed to treating their colonial subjects as equals, and this was the cause of mounting resentment and eventual rebellion.

In March of this year (1939) Mahatma Gandhi began a "fast unto death" to protest against autocratic British rule. War clouds were gathering over Europe. The most evil influence of Nazi Germany was sweeping relentlessly across Europe. Amid such worldwide turmoil my native Ceylon still remained an island of serene tranquillity and apparent content-ment, showing no cohesive signs of resistance to the British crown.

In Astronomy, 1939 was the year when a Swedish astronomer by name C. Schalen first showed that the universe contained vast amounts of cosmic dust (microscopic dust particles) that blocked out the light of dis-tant stars — a subject with which I was to become deeply involved in later years. The many conspicuous dark patches and striations that seemingly break up the whitish band of stars of the "Milky Way" had been known, and indeed catalogued for several decades. But these dark clouds were wrongly interpreted as being gaps and holes in the distribution of stars. Schalen was the first astronomer to convincingly prove that they were not holes, but dense clouds of dust particles that blocked out the light from background stars. His case for saying the particles may be made of a metal like iron turned out to be wrong, but it was with Schalen's work in 1939 that the subject of interstellar dust was born. Prophetically, 1939 was also the year that Fred Hoyle was elected to a Fellowship at St John's College Cambridge. As it will transpire later, it was with Fred Hoyle that my sci-entific career and my own story came to be inextricably linked.

Returning to my own story, my parents were of middle-class Sinhalese descent and their families had lived in or around Colombo for many gen-erations. My mother and my maternal grandparents belonged to the landed gentry of the city, whilst my father was descended from more rural stock in the southern provinces of the island. My paternal grandfather Dionysius Lionel Wickramasinghe worked in Colombo as a loyal servant of the British Crown, for which he was rewarded with a national honour bearing the title *Gate Mudaliyar*. He was conferred this honour by the Governor with a gift of a sword of which he was immensely proud.

My father Percival Herbert (PH) was born in 1909 and was educated at Royal College Colombo, a school modelled in the style of Eton and Harrow. My father's stint as a pupil at this school had verged on being legendary. He won every conceivable school prize, ranging from prizes for the Classics, Shakespeare to Mathematics and Science, and also a prize for the "Most Distinguished Student". Winning a scholarship from the British Colonial Government PH proceeded to Trinity College Cambridge in 1931 where he graduated with the highest distinction as a B Star Wrangler (first class honours) in Part 2B of the Mathematical Tripos in 1933 as a Senior Scholar at Trinity. The fact that he actually specialised in astronomy in his chosen topics for "Schedule B" (the advanced option), and that he had established a link with Cambridge astronomers of the time was to have a bearing in my own later career.

My early years of carefree life as a child were interrupted with events that still linger in my distant memory. Although the significance of these events eluded me at the time, what I remember vividly was the thrill of crawling in and out of an air-raid shelter that was built right in front of our house at 35 Hildon Place Bambalapitiya. I believe this must have been shortly after February 1942 when the Great War in South East Asia was raging and Singapore had just fallen to the Japanese. What Winston Churchill had called the "ignominious fall of Singapore" soon reverberated throughout the entire region with the conflict spreading to Burma and beyond. Trincomalee was the most strategic harbour in the area, and the Allied Forces considered the threat to Ceylon as being very real during the years 1942–1943. There were in fact several Japanese air raids that caused a moderate amount of damage to buildings and property on the Island, but where we lived in Colombo remained safe, and the strength of our personal air raid shelter was fortunately not tested.

Battalions of soldiers under South-East Asian command were posted throughout the island, particularly in a camp in the hill country in Diyatalawa, a place to which we often went for holidays in later years. The outline of a fox in flight was erected by these soldiers from large blocks of white limestone, thus modelling what is now the famous "fox hill" or "nariya kanda" in Diyatalawa.

For servicing the military effort in South East Asia, Ceylon played a crucial role in supplying much of the rubber that was required for various types of equipment. Rubber was a major economic crop of Ceylon at the time, and in order to meet the war targets, there was slaughter-tapping of rubber, severly damaging the country's rubber plantations.

Although World War II and the atrocities of Hitler were plaguing England and many countries of Western Europe they had much less impact on Ceylon compared with the war in South East Asia — the war against Imperial Japan. I was of course too young to grasp the horrific nature of the War on all sides, in particular the horrors resulting from the deployment of the atomic bomb over Hiroshima and Nagasaki. Not until much later when I had the opportunity to visit the Hiroshima War museum did I realise how barbaric an act this was.

After Japan's defeat in a brutal and savage war, the San Francisco Treaty of Peace in 1951 sought to curb Japan's military capacity and also asked for payment of reparations for the damage caused during the war. At this conference the leader of Ceylon's delegation J.R. Jayewardene delivered an eloquent speech opposing the demand for reparations.

> "We in Ceylon were fortunate not to be invaded, but the damage caused by air raids......etc....entitles us to ask that the damage so caused should be repaired. We do not intend to do so because we believe that in the words of the Great Teacher (Buddha) whose message has ennobled the lives of countless millions in Asia, that "hatred ceases not by hatred but by love". He ended his oration by saying "This treaty is as magnanimous as it is just to a defeated foe: We extend to Japan the hand of friendship and trust that with the closing of this chapter in the history of Man, the last page of which we write today, and with the beginning of the new one......her people and ours may march together into the future to enjoy the full dignity of human life in peace and prosperity."

I have dwelt on this gem of Sri Lankan political history for two reasons. Junius Jayewardene was my father's school friend, and at a much later date he was to enter my life when, as President of Sri Lanka, he invited me to be his advisor on science in 1980 and to set up the Institute of Fundamental Studies in Sri Lanka.

My own warmth of feeling towards modern, post-War Japan stems from the fact that I share with the Japanese an abiding respect for Buddhist values — precisely the sentiment expressed in Jayewardene's speech. Much later in my life I have had close working links with Japanese astronomers and also with an eminent Buddhist Scholar, President Daisaku Ikeda, of whom I shall have more to say at a later stage.

The country of my birth is steeped in Buddhist traditions. A century after the rule of Asoka the Great in India, Buddhism almost disappeared from mainland India, but spread overseas to Ceylon, Burma, Siam, China and Japan. Amongst these countries Ceylon (now Sri Lanka) is perhaps the most important centre of a form of Buddhism that was closest to the original form preached by Gautama Siddhartha in the 5th Century BC. The island is strewn with gigantic stone statues of the Buddha, temples and dagobas declaring its place as the home of one of the greatest religions of the world.

I was brought up as a child in the traditions of Buddhism. We had a shrine room in our home with a small clay statue of the Buddha, and we would make offerings of flowers to reaffirm our respect and allegiance to the Great Teacher. It could be said that the Buddha introduced a religion to the world that sought to make human beings less cruel to each other and less limited in their sympathy, even extending the bounds of sympathy across the whole spectrum of sentient life. My parents, despite their enlightened educational background, not only prided themselves as Buddhists in a formal sense, but also participated in the rituals which I myself later considered to be largely irrelevant.

During my early years, I went along with all the Buddhist rituals of poojas (worship) at shrines and regular visits to temples, uncritically, and I even remember enjoying in some way the dignity and solemnity of these occasions. I can recall most vividly those special occasions when Buddhist monks would be invited to the house for chanting *pirith* (Buddhist stanzas) throughout the night. They would sit in a circle of chairs draped with white cloth, and arranged within a makeshift tent made of palm fronds (a *mandapaya*). They would come in relays a few at a time, and we devotees would sit outside the tent on the floor, cross-legged, hands together in worship, and listen in silence. I still do not know the intended purpose of such rituals — perhaps they took place as blessings — but as a child I found them in some mysterious way deeply inspirational, and the monotonous sound of chanting soothing to the senses.

Uncritical adherence to such traditions follows when beliefs become instilled in childhood through force of habit and a desire to please and conform. With maturity comes self-criticism and a consequent need to reassess and reformulate the basic tenets by which one chooses to live. However, of all the religions that I have subsequently encountered, I find Buddhism even in its more formal style the most appealing, and perhaps expressing a still mysterious connection between our lives and the cosmos. A poem I penned at a much later date encapsulates this feeling:

"Full-moon night
The temple yard is a white sea
Of jostling pilgrims.

Shouts of "Sadhu! Sadhu!"
A stream of solemnity
Quietly flows
Into the shrine room
Brimming with the scent of temple flowers
And the fragrance of incense.

Pilgrims kneel in humble veneration:
Sadhu! Sadhu! Sadhu!
To thee we offer homage.

From aloft
A stone Buddha
In the dim light of flickering oil lamps
Smiles peacefully.

CHAPTER 2

THROUGH THE MISTS OF TIME

Through the hazy memories of childhood I can recall one occasion when, upon the warning of an imminent Japanese air raid, I was taken home in great haste from a Kindergarten classroom. This must have happened when I was about four years old and attending the Kindergarten at a neighbouring girls' school "St Paul's Milagiriya" which was just a short walk away from our house. I do not remember being overcome with any sense of fear or anxiety, but rather the thrill of creeping into the air raid shelter that was built just outside our house. As I mentioned in the last chapter such threats of air raids in Colombo came to nothing. The defences of the South-East Asia command in the region were a strong deterrent, and the nature of the terrain in the island was such as to prevent a Japanese invasion of the kind that happened in Singapore. The Japanese threat to Ceylon eventually receded over the following months until the time of their ultimate defeat in 1945.

What I remember of my first years of life are of necessity vague and imprecise. I recall how I used to sit in an open-plan class room at St Paul's Milagiriya overlooking green fields, and always clamouring to be in the front row of the class. I remember also that I was plagued by a fear of abandonment. What if no one was there when the bell rang to go home? What if both my parents had been killed in an air raid? Anxieties of the same type must of course be common to most kids of four, and I could (at a very much later date) see how similar fears plagued my own children and grandchildren to varying degrees. This type of fear must have a primeval origin going all the way back to our tribal ancestry. In primitive Palaeolithic communities the young would sometimes have to be left alone while the elders went hunting, and sometimes the parents were killed by tigers and did not return; and at other times the children themselves may have become

11

prey to predators. Such calamities would have been infrequent even in the historical past, and in modern times they are mercifully rarities in the extreme. Rationalising the patterns of such events and reordering our thoughts and fears according to a sober assessment of probability cannot be easy at an early age. But sooner or later, as night follows day, confidence in cycles of repeatability would allay our fears.

There are even greater fears that must haunt a young child when it comes to know that death must come to everyone. What happens when we die? What happens when we are locked up in a dark coffin? Do we feel pain when our bodies are burnt after death? I remember agonising over such questions at a very early age. Religions step in to offer answers to these questions and to placate our fears. These may take the form of promises of heaven or of eternal life. But the intelligent child will often not be content with such insubstantial offerings. In metaphysical matters we, unfortunately, do not have much factual evidence to guide us, and so we are encouraged to rely upon faith.

On the matter of life after death Buddhism does indeed offer an answer of sorts. We are reborn, either as a human or a lower animal depending on what we have done in our lives — depending, as they say, on our Karma. Karma is interpreted in general terms as a cause and effect cycle, which may have an *a priori* plausibility to some degree, if one relates it to physics. In any case the concept of Karma gives a sense of moral purpose. If you are a good and kind human being in the manner you conduct your present life, your next life will be as a better or more fortunate human being. If, on the other hand, you are cruel and do dreadful things, you could be reborn as a lower animal, even as an insect! These unproven propositions could have an effect on a young child that may either be good or bad. They could provide an incentive to do good things. On the other hand, they could lead to a total disbelief in an entire system of thought that might be seen to defy logic, or for which there is no empirical basis. In my case the effect was to instil a sense of fear, even terror at times. I would often dream of dying and being reborn as a hideous creature, or as a harmless insect.

The rebirth beliefs that I have described are not central to the main canon of Buddhist beliefs — exhorting the virtue of compassion and universal love. Gautama Siddhartha, the founder of Buddhism, was a Royal Prince who relinquished his princely life to seek answers to the difficult questions that concern our existence. He witnessed a surfeit of suffering all around him and sought to discover the cause of such suffering. The realisation or enlightenment he reached was perhaps an obvious truth — that the primary cause of suffering is attachment. If we can rid ourselves of attachments, we get rid of suffering.

Buddhism in its purest form is a highly personal quest for self-realisation, following the "noble eightfold path" — right thought, right intention, right speech, right action, right livelihood, right effort, right mindfulness and right concentration. This is extolled as the recipe to end all suffering and attain *Nirvana*. I imagine that Buddhism in its original form would have been free of ritual and ceremony. But with the passage of time the types of ritual that form an integral part of Hinduism must have crept back to become incorporated into the forms of Buddhism that now exist, and are practiced in countries like Sri Lanka.

The Buddhism to which I myself was exposed in childhood, as I said earlier, was full of ritual, with an excess of reverence being accorded to the clergy. My maternal grandfather Benjamin Soysa was a wealthy landlord who was a principal patron (*Dayakaya*) of one of the main Buddhist temples in Colombo, the *Vajirarama* Temple. It was most probably for this reason that our family was drawn to the ritualistic side of Buddhism. Full moon days (*Poya* days) are considered sacred for Buddhists and the temples are thronged with devotees clad in white. They lie prostrate before statues of the Buddha, offering incense and flowers and paying homage to the Great Teacher. One particular *Poya* day in the year — the full moon day of May — is more special and more sacred. It was on such a May *Poya* day that the Buddha is said to have been born, attained enlightenment, and also passed away, so this day is triply significant to Buddhists. It is marked by a colourful festival known as *Vesak*. In Buddhist countries like Sri Lanka Vesak is celebrated on a rather grand scale. Pandals adorn every street corner, depicting scenes from the life of Buddha, brightly illuminated with millions of electric lights. One would hear the monotonous chanting of *Pirith* — recitation of Buddhist stanzas in Pali — blaring out through loudspeakers throughout the city. The air is thick with the smell of incense.

Most Buddhist homes are illuminated with strings of coloured electric lights as well as home-made paper lanterns lit by the dimly flickering flames of candles. My early childhood recollections are of helping elders make such Vesak lanterns, with bamboo sticks tied up into octahedral frames, the faces of which are pasted over with coloured tissue paper. On every Vesak day such lanterns with candles flickering inside them would adorn the entrance to our house at No. 35. These same celebrations take place today in Sri Lanka, only on a much bigger scale and the spectacle has been modernised. The pandals are now illuminated with flashing neon lights, LEDs and strobe effects, and crowds of not only worshippers but also sightseers throng the streets of Colombo on this special day in May.

Returning to my own story, my privileged position as the only child in the family was changed with the arrival of my brother Sunitha in July 1942. Although my mother did not go to work, our upbringing in the early years of life was entrusted almost entirely to servants called *Ayahs*; and I remember that Sunitha and I each had our own *Ayahs* to look after us and prevent us from coming to any harm. Although I do not remember the exact details of our household at this stage, there were two other servants in our household — a woman to assist in the cooking, and a man who would drive the car, "the driver". We also probably had a rickshaw puller more or less at our disposal for short trips, for example to the temple, or school or perhaps the market. Rickshaws at the time were drawn by human effort. They consisted of an elevated carriage with two large wheels pulled along by a sinewy man, sweat dripping down his back. The large wheels of the rickshaw gave a considerable mechanical advantage and minimised effort to some degree, but the inherent cruelty of the system of slave-drawn transport cannot be denied. It was a relic of the slave system that persisted in colonial times. At the time, however, no one would have given a second thought to the grotesque inhumanity of riding in a rickshaw pulled by a human runner. Such man-drawn rickshaws continued in use in Colombo well into the early 1960's, being replaced by trishaws — runners replaced by men peddling a tricycle, and eventually by modern motorised three wheelers or Tuc-Tucs.

The nature of the social divisions, tantamount to a system of slavery that existed during the time I was growing up was a direct result of 300 years of colonial rule. Most empires that expanded through conquests have tended to regard subject races merely as a means to enhance their own wealth and prosperity. Although slavery in England had been abolished a long while ago, it was convenient to re-invent a similar system in the subjugated territories, and this happened not only in Ceylon, but throughout the British Empire.

In the 12th century AD, when Europe was in the Dark Ages, the Kingdom of King Dutugamunu of Ceylon was glorious, as is clearly evident in the magnificent palaces, gardens, temples and irrigation schemes that still survive in Anuradhapura and Polonnaruwa. The rule of Dutugamunu was said to have been a compassionate rule, guided by Buddhist principles and precepts. From these early days to the time of Portuguese conquest a largely benevolent feudal system prevailed throughout Ceylon. The dignity of life was respected through all strata of society within an economic structure that was largely based on agriculture. Farmers would have had their own small holdings which they cultivated, and they lived off the produce of their lands. The king or landlord levied taxes, but people were free to lead dignified lives. Besides agriculture other trades and industries also flourished. Mining of precious stones, fishing, cultivating spices including cinnamon, and tapping toddy from coconut palms were all separate occupations and trades. It is interesting to note that these separate trades gave rise to an elaborate caste system in Sri Lanka. There is a caste derived for example from farmers (Govigama), and another derived from toddy tappers (Durawa). For a long time these caste divisions remained an integral part of Sinhalese society, and marriages took place mostly within a single caste. My own family belonged to the Durawa caste — the caste of the toddy tappers. My father would jest that this was why some of us enjoyed our tipple — although he himself was a teetotal.

Whilst British rule sought to transform a largely self-reliant agricultural society into an urbanised commercial one, it served their ends to institute forms of slavery such as I have described. The colonial rule benefited Ceylon in some aspects, particularly with the construction of roads and railways and the institution of an administrative and educational system from which I myself (amongst millions of others) have benefited.

It served the interests of the British Empire to educate the most able and gifted Ceylonese and train them to assist in the running of the Colony. Schools such as the Royal College Colombo and University College, Colombo (the predecessor of the University of Ceylon) attached as an external college to London University served this purpose. My father was one who enjoyed the benefit of this educational system and emerged after a brilliant career at Royal College Colombo and Trinity College Cambridge to be selected to serve the British in their coveted Indian Civil Service.

Selection to the Indian Civil Service (ICS) was conducted by an examination that was reputedly the most difficult to pass. There was also a far less prestigious and comparatively easier Ceylon Civil Service examination, but in the year that my father graduated there were apparently no vacancies in this service. So it was that my father started his working life as an Indian Civil Servant with his first posting as the Deputy Collector of Customs in Bihar. This did not last long, however, because he felt so isolated in India and so homesick that in 1936, in less than 2 years, he was back in Ceylon. In 1937 he married my mother Theresa Elizabeth Soysa, and I was born on 20 January 1939.

Because an automatic transfer from the Indian Civil Service to the Ceylon Civil Service was not permitted he now had the difficult task to find an alternative employment. His first choice was the Professorship of Mathematics at University College Colombo, a post that just then had fallen vacant. My father applied for this vacancy with strong recommendations from distinguished Cambridge astronomers Sir Arthur Eddington and W.M. Smart who had taught him. But he was not successful in the competition, and I gathered later that this was a cause of much disappointment. Over the next few years he was forced to take up various types of administrative postings in the British Colonial government in Colombo, but none of these did he find entirely satisfactory.

In 1946 the post of Chief Valuer of Ceylon had fallen vacant and my father was appointed as the first Ceylonese to this post. Since his training in Mathematics was not exactly relevant to the new job, a condition of the appointment was that he be sent to London for two years to be trained at the University of London where he had to obtain a BSc degree in Estate Management. Insulting as this sounded to him (and to me in later years) it was considered at the time a necessary step to take in his journey of life. And it was this step that took our entire family to spend two years in London.

By now, in 1946, my younger brother, Dayal (who later became Professor of Mathematics in Canberra, and is also an astronomer) was a toddler of a year old. Along with him, Sunitha (now four) and my parents, I was to venture in my own journey of life to the hub of the British Empire.

CHAPTER 3

SAIL AWAY ON *SS* SCYNTHIA

The end of the long and hard-fought war against Nazi Germany brought in its wake a period of austerity and hardship for England. In 1946 London suffered from one of its worst recessions in many years, and there were acute shortages of foods and other essential commodities. Food rationing was in force, and there was a lot of visible poverty on the streets of London.

In the world at large historic events were beginning to unfold. The Iron Curtain, separating the Soviet Union from the rest of Europe was established, and the Cold War became a reality that was to last for many years. The Japanese Emperor's status as a demigod was at last downgraded. Plans for Indian independence were well under way, with Prime Minister Clement Atlee finalising the detailed arrangements for the transfer of power from London to Delhi. The partition of India, with all its attendant horrors, was also agreed.

In the world of technology the first images of the Earth from Space were obtained from rockets. In astronomy Fred Hoyle published a seminal paper in the *Monthly Notices of the Royal Astronomical Society* describing the processes by which carbon — the element of life — is synthesised in nuclear reactions taking place in the deep interiors of stars. It had been known for some years that the energy produced by stars is the result of the fusion of four nuclei of hydrogen to one of helium, and that this nuclear reaction would keep the sun shining for billions of years. What happens next, when all the hydrogen was finally converted to helium, had posed a problem that at this time was unresolved. Fred Hoyle solved the problem by showing how the element carbon will be the next to be produced from helium, and this groundbreaking discovery was eventually to transform our understanding of the universe. It also provided the reason for how living creatures came into existence, depending as we do on the chemical element carbon. This was to be of importance in my own astronomical journey.

Whilst all this was happening, in Cambridge England, across from the North Sea in the ancient University of Utrecht in Holland, an intrepid student, Henrick van de Hulst, was facing an austere group of distinguished astronomers to defend his PhD dissertation. This dissertation had argued that the dust particles in interstellar clouds were composed of ices that condensed out of the very tenuous gases that existed there. He passed the examination with honours, of course, and his monumental thesis on this subject became a milestone in the progress towards understanding the nature of cosmic dust, and indeed our origins, as we shall later see. At a much later date, in 1961, I had the pleasure of meeting van de Hulst which I shall touch on later.

The same year, 1946, also marked a milestone in my own journey of life. I had now finished two years of primary schooling at St Paul's Milagiriya before embarking on a voyage of a lifetime on board the Cunard White Star liner *SS Scynthia*, travelling from Colombo westwards to England. There can be no better education for a child of seven than travelling to new lands, and our voyage was no exception in this regard. Three weeks on a luxury ocean liner sailing on the high seas was itself a transformative experience, and the countries and places we visited on this voyage etched a deep and lasting impression on my mind.

The urge that drives every great explorer to discover new lands must to some degree be present in everyone of us, and this would perhaps be particularly strong in our earliest years of life. To approach the world with fresh unjaded eyes and make judgments for ourselves is the happy prerogative of childhood which is all too quickly lost with the introduction of formal education. Most prejudices particularly in relation to people and places are taught, and some are obviously more harmful than others. We are not born racists or misogynists but too many of us are taught to become so.

I remember our ship *SS Scynthia* sailing through the Suez Canal and the many strange and wonderful sights that greeted us on the way. It brought to life stories from the *Arabian Nights* that I remembered from very early in my childhood, and the time when I acted in a school play as *Ali Baba*. Egyptian vendors clambering onto the decks of the ship on ropes, selling a variety of goods and our purchase of one particular item, a child's metal folding deck chair, are some irrelevancies that have stuck indelibly in my memory. I remember also my first sight of camels and being told that their humps stored water to equip them to be "ships of the desert". There is no doubt that travelling widens our perspective of a world that is wondrously varied.

Over a period of hundreds of millions of years the surface of our planet has been shaped and moulded. Plate movements, volcanic eruptions and recurrent episodes of mountain building have carved out the contours of our planet, moving our continents and changing the distribution of life — fauna, flora, animals and humans. A journey across the world also cuts through the timeline of history, the story of human migrations, civilization, conquests and empires over periods of thousands of years.

Arriving in England on board *SS Scynthia* crossed such a timeline. Compared with my native Ceylon of the 1940's England looked like an alien world. Everything was on a much larger scale compared with my home country — the streets were wider and more crowded with cars, taxis and red double-decker buses. The buildings were magnificent in their size and grandeur compared with anything one could see in the Colombo that I left. And to a boy of seven the underground system, with tube trains whizzing through a network of tunnels, must have looked like a mythical wonderland. I also remember meticulously figuring out routes on the London Underground maps. And I can still recall my fascination with the London double-decker buses, and particularly the neat rectangular cardboard tickets torn out from blocks carried by ticket conductors, and their "any more tickets please" invocations. We had not been on public transport before and it seemed infinitely more exciting to us children than our chauffeur driven car at home.

Our first floor flat in Hampstead at 23 Pattison Road was just a short walk away from Hampstead Heath. Among the trivia of daily life that have stuck indelibly in my memory was the milkman delivering milk bottles sealed with silver and gold tops, and particularly the bottles of yoghurt, which I liked! I recall attending the nearby primary school, but I remember far less of my time at school than I do of life outside school. At weekends I looked forward to long walks on Hampstead Heath with my parents and the freedom to roam around its meadows and coppices. In those days children would have far more freedom to roam around a park with safety, and instances of crimes against children or abductions were exceedingly rare.

Despite the tough times that most people had to face in post-war England, one did not hear of the various kinds of heinous crimes that are so common today. For me, Hampstead Heath was a pristine island of tranquillity and beauty in the midst of a bustling, if not somewhat disturbing, conurbation that was London. The differences in the vegetation and bird life, fauna and flora, from that of tropical Ceylon would have been easy to spot even for a child, so also the changing patterns of life on the Heath through the cycle of the seasons — Spring, Summer, Autumn, Winter. I also remember being told for the first time by my father about how the seasons occur. A toy globe with an inclined axis moved around a central light illustrating the seasons was my first introduction to an astronomical discovery.

To anyone used to the tropics with barely any seasonal changes through the year, the change of seasons must come as a surprise. The first snow that fell in the winter of 1946 transported me to yet another wonderland. It felt as unreal as walking into a story book landscape that had escaped the bounds of fiction and suddenly came to life. I had of course seen images of snow scenes on Christmas cards and in books, but never thought to connect them to any events or places in the real world.

My abiding memories of this time are of my father wearing a thick woollen (Trinity College) muffler, a thick woollen coat and gloves and taking us for long walks in the snow. There was a great deal of snow in the winter of 1946, which had in fact been the coldest on record for a decade or more. The "big freeze" of this winter continued well into the new year and I remember the fun I had, particularly with my brother Sunitha, making snowmen and sliding down snow-clad slopes in a toboggan.

With post-war stringencies in force, coal and fuel were rationed and we would have had difficulty keeping as warm as we would like to have been. But I do not recall having any real cause to complain; we coped with the cold by wearing extra layers of woollen clothing. Nor did the food rationing that was in force cause our family any hardship. But I dare say there would have been very many others who did suffer, for this was the time of a great depression, with levels of poverty in London that had not been seen for a long time.

Although people from a tropical country will not take kindly to the bitterly cold weather we experienced (except perhaps as a novelty for a brief period) the indigenous people of the British Isles must, I think, be genetically equipped to deal with such conditions. They are mostly descended from Nordic tribes that lived on the edge of ice sheets during the last ice age 10,000 years ago, eking out a precarious existence under bitterly cold Arctic conditions. Such circumstances would have moulded the constitution of these people, whilst also providing a stimulus to creativity of the kind that would be needed for devising strategies for survival under these adverse climatic conditions.

In the 1940's, short-term variations of climate, with occasional sharp discontinuities such as those witnessed in 1946, would have been regarded as part and parcel of a natural cycle. Climate change ideas had not yet become as established as they have been in recent years. In a blame culture that prevails today we are always ready to impugn this or that sector of society for whatever ills that befall us. Such behaviour is perhaps also the relic of a primitive instinct derived from the days of our primal ancestry when we attributed all our ills to devils or magic or witchcraft! Climate change, which is a big political issue nowadays, is widely blamed *exclusively* on human activities, although evidence for this still remains equivocal. The issue is not whether there *is* global warming, which is of course a fact, but how much of it is due to man-made CO_2 emissions. Oscillations of the average temperature of our planet have undeniably occurred over timescales of centuries and millennia. There was, for instance, a warm period during Roman times in England when grapes were cultivated; and there was a mini-ice age in the seventeenth and eighteenth centuries, and all this could be part of a natural cycle during a period when no man-made CO_2 emissions could have played a role. Human activities have undeniably made a contribution to global warming in recent decades, but the question is, how much. Politics unfortunately intervenes and plays a more important role than science in issues of this kind.

My own keen interest in natural phenomena may have had its earliest beginnings in the cold winter of 1946, when we were all huddled together as a family in our small London flat. I remember asking my father so many questions about the natural world and being impressed by the way in which he was able to explain even the most complex ideas in a simple fashion. If he was given the chance he would, I think, have made an excellent teacher and scientist. I also had a chance at this time to discover things for myself, to wander alone in nearby parks observing Nature, counting the numbers of petals in flowers and noticing patterns, watching also the patterns formed by clouds in the sky that always intrigued me.

I also remember my first Christmas in London and what I got from Santa Claus! It was my very first Science Kit (an electricity set) that I prized most of all and of all things a few bananas — all packed into a large Christmas stocking! Bananas were a scarcity then, and it being so were considered a luxury. It was, however, my electricity set that I remember so vividly to this day, and the excitement I had with connecting batteries, circuits, bulbs, an electric bell and switches. I sometimes think that it was these experiments that laid the foundation for my later professional interest in electromagnetic theory, the theory I needed in order to probe the behaviour of cosmic dust. There were of course no transistors or solid state devices of any kind in use in those days. In fact it was in December of this very year, 1947, that the "point contact transistor" was discovered at a research level. Nine years later in 1956 John Bardeen and Walter Brattain were awarded the Nobel Prize for Physics for the discovery of the transistor.

Aside from playing with science kits I also spent much time in long winter evenings reading books. From an early age I preferred factual books to fiction, but there were two story books I remember reading: Swift's "Gulliver's Travels" and Kingsley's "Water Babies" most probably in abridged form. These bizarre and surreal fantasies captured my imagination and led me to read other story books leading eventually to "Lamb's Tales from Shakespeare". The fiction that attracts children must surely be linked to the times in which they live. Nowadays children seem to go in for high-tech adventure stories involving black holes and time travel and there is also an obsession with aliens. The latter might well be connected with the prospect of discovering alien life in the not too distant future, and possibly even making contact with alien intelligence!

Apart from fiction, poetry and particularly memorising and reciting poems aloud became a passion from an early age. I loved the inherent lyricism of the English language, so reading poems contained in anthologies such as "A Children's Garland of Verse" was an enjoyable pastime from a very early age. I also began to experiment in writing my own poems and this too was an activity I enjoyed.

To conclude this chapter I should say something about the level of discipline to which I was subjected as a child. I have seen attitudes of parents that range from total indulgence to rigidly enforced rigour, and these different regimes may suit widely varying temperaments of children. In my own case my parents were lenient almost to a fault. One possible reason for this excessive leniency, which of course I did not regret, was that both my parents were themselves brought up in a regime of discipline that verged, it seemed, on tyranny! So they may perhaps have sought to rectify this in the upbringing of their own children.

Through my childhood I felt that I was given the freedom to develop my personality within limits that were set more by example than by force. There was no incentive to disobedience of the implicit rules that were thus defined by example; and I believe that my conduct as a child was exemplary both at home and at school. It is in my view an unwise parent who has the habit of always saying "don't do that" without explaining to a child why "that" would do any harm.

CHAPTER 4

FROM SCHOOL TO UNIVERSITY

When I left Ceylon with my parents in 1946, the island was still a British Colony administered from Westminster. When I returned two years later it had been granted independence to become an independent Dominion of the British Commonwealth, still retaining the King of England, King George VI as its statutory head of state. Following independence, however, Ceylon was effectively free to govern itself through its locally elected legislature. A parliamentary system modelled on Westminster had an elected lower house, the Parliament, as well as a nominated upper house, the Senate. The first Prime Minister of independent Ceylon was D.S. Senanayake, and the first Governor General representing the King was Sir Henry Monk Mason Moore. The Dominion status of Ceylon continued until 1972 when it became the Socialist Democratic Republic of Sri Lanka, thereafter retaining only a nominal link as a member state of the British Commonwealth of Nations.

From 1948 for the next 12 years I remained in my native country, growing up in an anglicised home, but with a cultural and religious backdrop that was distinctly Ceylonese and Buddhist. Throughout this period I enjoyed the cosy comforts and security of a happy family life at 35 Hildon Place. My parents, three brothers and I formed what was very much an isolated family unit with little or no outside contacts. We enjoyed our regular family holidays together, to the hill country — Diyatalawa and Nuwara Eliya in particular — where we would rent government bungalows and spend a couple of weeks at a time during the school holidays. Our daily routine would be to take long walks in the hills and come back to sumptuous hot dinners prepared for us by the "bungalow cook". The chilly evenings were spent by a log fire either reading books or playing games. Such holidays are among the happiest memories of my childhood. The train journeys to these resorts in old steam trains chuffing up the hills along winding railroads, the smell of coal, the luxury of first class sleepers, equipped with hand-wash basins and ensuite toilets, remain to this day a vivid memory.

Back at home at 35 Hildon Place, I shared a room with my brother Sunitha in my early years, and we engaged in many hours of serious, and not-so-serious, conversations. When Sunitha's school interests diverged from mine — mine in maths and physics, his in biology — we had less serious stuff to talk about, but I still remember taking more than a passing interest in his experiments in biology, dissections of animals — cockroaches, frogs — and worrying sometimes about the conflict of such activity with our Buddhist beliefs. I think my fascination for biology, that was to reach maturation at a much later date, may have arisen through these early encounters with Sunitha. When, at a much later stage, Sunitha had become a distinguished biologist (he was Professor of Haematology at Imperial College, London) I still enjoyed talking with him on matters scientific — even Astrobiology — on the occasions when we met.

With Dayal, who was some six years my junior, I think I must have felt a more protective and possibly even inspirational connection. He was unfortunately the victim of poliomyelitis early in his childhood, and had difficulty walking. Despite this handicap he excelled in his schoolwork, particularly in Mathematics, and it was my chance eventually to encourage and stimulate this interest. This I believe I must have done, at least to the extent of influencing his decision to come to Jesus College, Cambridge to read mathematics, and later do a PhD in astronomy. Needless to say he excelled in this activity and later became a distinguished researcher in astronomy and a Professor of Mathematics at the ANU in Canberra.

I remember my youngest brother Kumar only as a kid. When I left for Cambridge in 1960 he was barely eleven. My recollections of him were as a boisterous and playful boy engaged in pranks and with little interest in his studies at this stage. But to the surprise of everyone he has ended up as a distinguished physicist and is now a Professor at the University of California at Irvine. He is in the forefront of research in the new disciplines of Nanotechnology and Nanomeasurement.

During our time in Colombo we did not have very much contact with our wider families as was the common custom. Although we did indeed have many cousins (my father had 5 brothers) our contact with them remained rare. We did, however, visit our two sets of grandparents on a regular basis. We would often drive to their houses in the evenings, sit out on their verandas, talk, or say nothing, and listen to the chirping of crickets and cicadas interrupting the stillness of the night.

Let me return now to the time line of my personal story. After returning to Ceylon at age nine, I spent nearly two years at Royal Primary School, a Preparatory School for Royal College Colombo. It was during these two years that I began to be an earnest student taking a serious interest in formal schoolwork. My passion both for mathematics and for science was slowly developing. I took my earlier interest in doing science experiments much further by reading chemistry books, acquiring kits for chemistry including elaborate components like Bell jars, Bunsen burners, glass tubing, flasks, pipettes — and of course lots of chemicals. The pantry at 35 Hildon Place was often taken over by me for my chemistry experiments, much to the chagrin of my mother. Besides chemistry I remember also my fascination with making crystal radio sets, with long aerials that had to be set up in the garden by tying wires across tall trees. I found it intriguing to use just an aerial, a tuning coil, a galena (mineral) crystal and a good pair of earphones to receive broadcasts from local radio stations. The BBC world service was often relayed by the local radio station (Radio Ceylon) and I remember straining my ears through crackling earphones to listen "the news read by Alvar Lidell". Besides crystal sets there were also my experiments with telephony, rediscovering Graham Bell! I cannot help feeling that all these early and spontaneous adventures of childhood established my lifelong fascination for science.

By the time I was about ten I was reading a wide range of books — from Sherlock Holmes, to Homeric Legends, and also avidly pouring through children's encyclopaedias learning facts about science. All my reading was in English, and I must admit, somewhat to my shame, that I rather neglected my mother tongue Sinhala, which unfortunately, was not spoken very much in our highly anglicised home.

I believe that such activities at an early age — experiments with science as well as reading, were possible only because there were no distractions. There was no television, nor computers or electronic games. Children in today's world are overwhelmed with such distractions. There is a tendency to acquire knowledge and information from computers and the hands-on use of a variety of modern technologies. The loss of reading time is in my view to be regretted, but one can wonder whether the new types of skills and knowledge that these modern activities can impart would have a role to play in the technological world of the not-so-distant future. Perhaps reading and imbibing knowledge from books would become less relevant to our future technological development and survival. I would hope this will not be so. But the future of technology and indeed the future of society is hard to predict. In the year 1949 no one would have dreamt of the technologies that simultaneously enrich and plague our lives today — computers, cell phones, instant worldwide communication, to name but a few.

At the age of eleven I had to sit a competitive examination to be admitted to Royal College Colombo. I entered Royal in 1950 and the following seven years that I spent at this school laid the foundation for my later intellectual development. Royal College, as I remarked earlier, was the school that my father had attended in the 1920's, and in the history of which he had left an indelible mark. On the many panels of prize winners that adorn the school hall the name of P. H. Wickramasinghe (1927) can be seen, and for prizes in an extraordinarily wide range of subjects — Shakespeare, Mathematics, Science. His almost legendary status in the history of this school was something I had to aspire to, or so I felt at the time.

This school was started by the British in 1838 some 20 years after the island had become a Colony. It was probably intended to serve the needs of families of British Colonial Servants who were posted here, and it was modelled very closely on the English Public School system. As time went by admission to the school was extended to include sons of rich and influential Ceylonese, and there was a distinction between the academic pupils and the ones whose merit was their wealth.

By the time I entered Royal College, the last in a long line of distinguished British Principals of the school had retired, and the first Ceylonese Principal J.C.A. Corea was in post. All the boys at this time were native Ceylonese, many of them sons of old boys of the school. The school can boast of a long line of distinguished alumni, amongst them Governors of the island, Prime Ministers, High Court Judges, Civil Servants and some scientists and academics. Just as in my father's case their names adorn the panels that listed prize winners and scholars down the ages.

My seven years at this school were perhaps the most important in my life. The school's emphasis on a high standard of scholarship is encapsulated in its motto "*disce aut discede*" (learn or depart) and in the school song which contain the lines "We will learn of books and men and learn to play the game". Compared with modern schools in the UK, there was much emphasis on discipline with draconian punishments, even canings, meted out for the most severe breaches of discipline. Like all public schools, Royal College also emphasised physical training and sports, but in this latter activity I did not excel. At times when there was "drill" on the timetable (in the first two years) I remember taking letters of excuse in which my parents had to perjure themselves to say that I was suffering from this or that intangible illness.

Whilst I found PT and Drill rather boring and pointless, a type of physical activity I enjoyed was practical work at the School Farm. This was a fortnightly timetabled activity in the first three years of Royal where the entire class was taken in a bus to the Royal College Farm situated in Narahenpita, in the outskirts of Colombo. Here we would have a whole morning of hands-on experience of tilling the soil, sowing seeds as well as tending farm animals. In the afternoon there would be a formal classroom session in the adjoining building where we would be taught the theory of farming and agriculture. I remember these days so vividly that the experience on the farm must have had a beneficial effect on me. It certainly enabled me from an early age to connect the food we eat with the toil that goes into its production. Planners of modern or post-modern school curricula may do well to try and bridge an important educational gap that exists at present in this regard.

I became aware of current issues in science almost from the start of my time at Royal College (age eleven). The *Ceylon Daily News* — the local newspaper — appears to have had a good coverage of science news in those days. I diligently kept a scrap book of the science news that inspired me, and my cuttings around December 1949 become almost obsessive about Albert Einstein. A cutting for December 29, 1949 (against which I have made many pencil notes!) reads:

"Professor Albert Einstein who attained fame by this Theory of Relativity and whose theory of Mass and Energy led up to the manufacture of the atom bomb has described today that he has completed a new theory that may, in time, unlock such secrets of nature as to what makes the Universe tick.

He calls the new theory a "generalised theory of gravitation" and it is designed to bring together under one understandable formula all known physical phenomena...."

As time went by my interests at school became more focussed on just a few subjects — Science and Mathematics. In Mathematics I remember reading widely and learning topics that were well outside the school syllabus. I remember also being an earnest student from the outset, but in my first three years the end-of-year reports, based on performance at exams, were a little disappointing. I was ranked 2nd or 3rd, not first as I felt I should have been! I remember also being told by the Vice-Principal M. M. Kulasekaran, (who had taught my father many years earlier), that this was not good enough for PH's son!

At this point I should say that there were extraneous forces at work that were difficult to overcome both in my school life and in Ceylon more generally. Tensions between the Sinhalese and the Tamils that were dormant for a long time were slowly beginning to surface. The boys who were ranked first and second were always Tamils, and the teacher who marked the exams was also a Tamil. So it appeared that a subtle form of favouritism and discrimination had worked against me!

The genesis of the Sinhalese-Tamil problem in Ceylon goes back to the days of the British Raj. In the early part of the 20th century there was no apparent disharmony between the two main ethnic groups of the Island — the Tamils of Dravidian origin, and the Sinhalese of alleged Aryan descent. The Tamils, who made up 15% of the population were, however, disproportionately represented in the British administration of Ceylon through a deliberate policy that favoured the minority race. Missionary schools, that had been set up in the Tamil-dominated North of the island, offered a much higher standard of education than elsewhere, and so provided the British administration with well-educated Tamil youth who became their chosen administrators and educators. This tipped the numerical balance in the British administration of Ceylon in favour of the Tamils, significantly outweighing their 15% share of the total population of the island.

In the run-up to independence in 1948 Tamil politicians had fought hard to ensure the continuation of Tamil dominance by demanding a form of constitution in which they had a 50% presence in the legislature. This demand was rejected by the British government and the elected parliament of independent Ceylon was constituted fairly and with no artificially enforced communal bias. In the event the 15% Tamil population of Ceylon found themselves with a 15%, rather than a 50%, share in the administration of the island. This is the main cause of the communal tension and strife that was to follow in later years.

Although my competitive instincts were challenged by the discrimination of being unfairly marked down to second or third place, as I described, I do not remember bearing any personal grievance or animosity against any of the favoured boys. In fact most of my friends were Tamils and the boy who always beat me was a particular friend. I am grateful to him (T. Devendran) for introducing me to a Tamil classic called *Thirukkural* which is a collection of rhyming couplets composed in the 1st century AD, and of which Albert Schweitzer said

"There hardly exists in the literature of the world a collection of maxims in which we find such lofty wisdom... ."

A few examples of these rhyming couplets (not always rhyming in translation) would suffice to convey what he meant:

"Tis never good to let the thought of good things done thee pass away;
Of things not good, 'tis good to rid thy memory that very day."

"Learning is the true imperishable wealth;
All other things are not wealth."

The Buddhist text *Dhammapada*, to which I was introduced through my visits to the Temple during my childhood, provided me with similar complimentary gems of wisdom:

"A man is not called wise because he talks and talks again; but if he is peaceful, loving and fearless, then he is in truth wise."

"The one who has conquered himself is a far greater hero than he who has defeated a thousand times a thousand men."

The *Dhammapada* together with the *Thirukkaral* gave me a framework of moral guidance and inspiration in my adolescent years.

I said earlier about the inherent lyricism of the English language and my interest in poetry. I believe that my passion for English poetry was inspired to a large extent by my father's somewhat strange habit of reciting poems aloud to himself. H.W. Longfellow's "A Psalm of life" still rings in my head from these very early days:

> "Tell me not, in mournful numbers,
> Life is but an empty dream! —
> For the soul is dead that slumbers
> And things are not quite what they seem...
>
> Lives of great men all remind us
> We can make our lives sublime,
> And departing, leave behind us
> Footprints on the sands of time."

> — H.W. Longfellow (1807–1882)

Besides Devendran, whom I regarded more as a friend rather than a classroom rival, I had few other friends in school. I did not take part in any sporting activity at school, so all the friendships I formed were in a classroom context with boys who, like Devendran, were in some way my academic rivals, or so it seemed. One exception to this was my friend Ponnambalam Rajendram (Raj), a boxer and rugby player, who was an indifferent scholar, and so could not be perceived as a rival. We spent much time together in the last two years at school, but went our separate ways after entering University in Colombo — he to Engineering and me to Mathematics. Later, however, our paths crossed in Cambridge, and still later in London, where he now lives with his wife Radha and daughter.

Returning to the timeline of my narrative, I continued to enjoy the processes of learning and acquisition of knowledge of all sorts, well outside mathematics and science. In my penultimate year at school (Lower VI) I remember acquiring a copy of Jowett's translation of the *Dialogues of Plato* and reading these dialogues made a deep and lasting impression.

I had thought at the time that dialogues of the Platonic type were the only reliable way to obtaining a knowledge of the true nature of things. Another book that made an equally strong impression, and one that I have read and re-read many times, was H.G. Wells' *A Short History of the World*. This book drives home most emphatically the fleeting and transitory nature of our place in the scheme of things.

I also read more English poetry now, exhausting the romantic poets and then moving on to the war poets, T.S. Elliot and the modern poets. I also took to writing poetry more seriously than I had done earlier, an activity that enabled me to vent my adolescent feelings on a wide range of subjects. A style of writing with which I found a personal resonance was "imagism", a form introduced into English poetry by the American Poet Ezra Pound. Ezra Pound was in turn influenced by the Japanese *Haiku* style of poetry. This led me to acquire many tomes of Japanese *Haiku* poems in English translation, including poems by the great Japanese poet Basho. The precise elegance and power of the Haiku style, comprised of short 7 syllable, 3 line poems, is possibly difficult to capture in English translations. The importance within this style to focus on a single moment of deep insight appealed to me. *Haiku* seems to me to have a precision akin in nature to mathematics.

Haiku started off as court poetry in the 9th century AD and resonates well with the Buddhist outlook on life. Both Haiku poetry and Buddhism dwell on the transient nature of our existence. My own early experiments with the *Haiku* encapsulated the essence of this style:

> Raindrops
> Cling to the leaves
> And fall reluctantly
> Drop, by drop

> The street lamp with its arched neck
> Peers
> Into a beggar's empty bowl

> The evening is silent
> Even the flowers fall noiselessly

My father had an extensive library and had a daily delivery of newspapers and I would read anything and everything I could get my hands on. I would often take notes if I found material that was of particular interest.

In July 1955 I had read in the *Ceylon Daily News* the headline "Mankind must renounce war or else perish." This was of course the famous Russell–Einstein denunciation of nuclear war, following the A-Bomb tests in Bikini, signed by Nobel Laureates from many different countries. The powerful rhetoric and eloquence of this speech still resonates with me:

> "We are speaking on this occasion not as members of this or that nation, continent or creed, but as human beings, members of the species man whose continued existence is in doubt... ."

After describing the gruesome horrors of nuclear warfare the declaration ended thus:

> "For countless ages the sun rose and set, the moon waxed and waned, the stars shone in the night, but it was only with the coming of Man that these things were understood. In the great world of astronomy and in the little world of the atom, Man has unveiled secrets which might have been thought undiscoverable. In art and literature and religion some men have shown a sublimity of feeling which makes the species worth preserving. Is all this to end in trivial horror, because so few are able to think of man rather than of this or that group of men? Is our race so destitute of wisdom, so incapable of impartial love, so blind even to the simplest doctrines of self-preservation that the last proof of its silly cleverness is to be the extermination of all life on our planet?........."

As a teenager with my life ahead of me I found this speech moving and deeply inspirational. Reading it I became more determined than ever to explore the "great world of astronomy" for myself.

From my position No 2 or 3 in class in the first three forms at school, I had now, in Form V, soared to first place, head and shoulders above my nearest rivals. My mathematics teacher at this time was Mr. Elmore de Bruin, a tall, fair-skinned man of Dutch descent, and an excellent and inspiring teacher of mathematics. Mr. Bruin regarded me as one of his star pupils, and it was under his able guidance that I began to appreciate fully the nature of "proof" in mathematics, its logical infallibility, and above all, its exquisite elegance and beauty.

Problem solving is of course the ultimate goal of mathematics — at any rate of applied mathematics — and there was nothing I enjoyed more at this stage than grappling with the trickiest of problems. I recall attempting to tackle questions set in the Mathematics papers for the Cambridge University Entrance Examination that my father had ordered for me. I had to struggle with some of the problems whilst I found others too easy. These exercises made me well-equipped to compete at an early age for a coveted Mathematics school prize — the Ruby Andries Memorial Prize for Mathematics, the examination for which was traditionally set and marked by the Professor of Mathematics at Colombo University. Not surprisingly perhaps, I won this prize and my name went up on the relevant Prize Panel in the school hall 28 years below my father's! I had also competed in the same year for the school's Shakespeare Prize, but I failed in this particular competition, being justifiably beaten to it by L.W. Athulathmudali, an Arts student who later became President of the Oxford Union, and much later a Minister of State in the Sri Lankan Government. Sadly Athulathmudali, who may well have become a future President of Sri Lanka, was assassinated by a Tamil Tiger terrorist in 2004.

Although my passion for mathematics should have directed me only to one course of University studies, I was for a short while tormented by a perception that I may prefer to read English rather than Maths, or even do a joint English and Maths degree at University. When I approached the Principal of the School (Mr. Dudley K.G. de Silva) to help me resolve this dilemma, I remember that he was dismissive to the point of being rude. I was thus left with no option but to continue with Maths alone. I was annoyed at the time, but now I feel grateful to him for his wisdom in directing me along the best route.

My serious interests in astronomy were also beginning to take shape more definitely at about this time. A benefit of living in Ceylon in the 1950's was that the environment was still pristine and unpolluted. As I have already mentioned there were no bright street lights in the suburb where we lived and hardly any pollution from cars or buses, so that the pageant of the night sky was magnificently brilliant. We lived close to a beach by which a rail road ran connecting Colombo with smaller cities in the south. In my teenage years I would often walk along this beach in the evenings, sometimes along the railroad sleepers, and watch the sun set over the Indian Ocean. I vividly recall the experience of spectacular sunsets such as I have never since seen. Being so close to the equator there is no extended period of twilight. The brilliant spectacle of a sunset merges in minutes into a black canopy overhead studded with millions of stars. Looking up at the myriads of stars that populate the Milky Way, contemplations about man's place in the universe were inevitable.

It is rare to see such a spectacle nowadays in our modern cities with their deplorable levels of light pollution. Our inability to enjoy this natural beauty of the night sky leaves us the poorer by far, and also less able to make the connection between ourselves and the wider cosmos — a connection that would have been deeply felt by our distant ancestors.

Ceylon was and still is steeped in Buddhist traditions, and the influence of Buddhism in daily life is inescapable. Buddhist descriptions of cosmology that date back to the early Christian era are distinctly post-Copernican. In a Buddhist text *Visuddimagga* (written in Sri Lanka in the 1st century AD) it is stated that:

"… as far the these suns and moons revolve shining and shedding their light in space, so far extends the thousand-fold universe. In it are thousands of suns, thousands of moons … thousands of *Jambudipas*, thousands of *Aparagoyanas*…"

the latter being translated as meaning extraterrestrial abodes of life. The billions of galaxies of modern astronomy could be identified in statements found in other contemporary Buddhist texts which referred to the entire Universe as "this world of a million, million world systems". Such passages made a significant impact on me in my young years, as I noticed a striking similarity between these ideas and those expressed by James Jeans in his *Mysterious Universe*.

My resolve to study astronomy was strengthened by a serendipitous astronomical event that was connected with my homeland. A total eclipse of the Sun, visible from Ceylon, was to take place on June 5, 1955. Ceylon which had hitherto been a scientific backwater, was suddenly transformed into a hive of professional scientific activity. This particular eclipse was to have the longest period of totality since AD 699 and several important scientific experiments were being planned. Scientists from Britain, USA, France, Germany, and Japan all converged here and the local newspapers were full of news about this momentous scientific event. One experiment that was planned was a test of Einstein's Theory of General Relativity which predicts a bending of light by a small predictable amount (1.75 arc sec) as the light of a star passes close to a massive object like the sun. The project was designed to validate an experiment of a similar kind carried out by a team led by Eddington during the solar eclipse of 1919.

As a keen amateur photographer, I had set up my own experiment with a simple camera fixed at the end of a home-built telescope to capture the event. We were of course warned about the dangers of looking directly at the Sun, so at the appointed hour, we had our darkened glasses and basins of water in place to watch the progress of the eclipse. I watched with bated breath as the Moon slid ominously over the Sun's disc, casting an instant gloom over the landscape as in an impending thunderstorm. Then total darkness descended suddenly, lasting for what seemed an interminable seven minutes. There was a noticeable chill in the air. Lotuses in our pond at No. 35 began to fold their petals, animals cowered and crows cawed wildly. The fabled spectacle of the solar corona with its outstretching flaming tongues was visible intermittently through transient clearings in a thin veil of drifting cloud. Then it was all over, the noon day sun mystically reappeared. I felt more than ever before the indomitable power of the cosmos.

As often happens with observations in astronomy most of the observing teams in 1955 were disappointed because clouds intervened, but a few stations were able to make successful observations that led to new science. To a teenage admirer of science these events provided a thrilling experience. Science was happening at my very doorstep, and the events of 1955 played no small part in my determination to pursue astronomy as a career.

Nowadays most students enter astronomy through undergraduate courses in physics, or physics and astronomy. When I asked around for advice on how one became an astronomer the answer I was given by informed persons, not least my father, was through mathematics. And this was the route on which I had already decided to embark. For my A-level examination, that was the selection test for the University of Ceylon, I took only three subjects — Pure Mathematics, Applied Mathematics and Physics. I not only passed all three subjects with distinction, but in the entire A-level examination of that year I was placed first in the island. I won a number of school prizes in my final year for my performance at the examination, and I entered the University of Ceylon in 1957 as an Entrance Scholar in Mathematics. Because of the excellence of my examination results I was also exempted from doing a first year course, and went straight into the second year specialising in Mathematics from the outset. Already at this time I had set my sights on my father's old University, Cambridge, should things work out for me as I had hoped.

Before I began my studies at University, political events in Ceylon were beginning to take an unfortunate turn. The highly erudite S.W.R.D. Bandaranaike, an ex-President of the Oxford Union, and an eloquent orator, was Prime Minister and he passed the infamous "Sinhala Only Act" in Parliament in 1956, an Act by which the official language of the island was to be changed from English to Sinhala (the language of the majority Sinhalese). This step led to ethnic riots in the Island whereupon the Act was rescinded and in a new Act of Parliament both Sinhala and Tamil were given parity as official languages on equal term. But by now the damage was done. Tamils began to feel alienated by the government of the island, and the seeds of the Tamil Tiger rebellion were sown.

The year 1957 was also memorable for another less parochial reason. In November of this year the Russians sent the first living creature — the dog *Laika* — to orbit the Earth aboard the space ship *Sputnik 2*. I remember this news as vividly as though it were yesterday, and of thinking of the cruelty involved in letting poor *Laika* languish and die in the depths of space.

The transition from my school life to University life in 1957 was an easy one because I attended University from home. The University of Ceylon, Colombo, was only 20 minutes away by bicycle or ten minutes by car. I took the three years of University studies in my stride, following most of the courses in applied mathematics, because I felt this was the most important tool for exploring the Universe. I was again lucky to have some excellent teachers who inspired me. A person who influenced me greatly in those early years was the professor of mathematics C. J. Eliezer who was himself a distinguished Cambridge product, a former Fellow of Christ's

College and a pupil of the illustrious physicist Paul Dirac, whose major success was to reconcile special relativity and quantum mechanics. Through Eliezer's lectures I obtained exciting insights into the theory of electromagnetism, a subject in which I was later to specialise. I did not realise at the time that Eliezer and Fred Hoyle (who was to be my mentor much later) were Cambridge contemporaries and that they both had associations with Dirac. Because of the connection between these three people — Dirac, Eliezer and Hoyle — it turned out by a really curious coincidence that Hoyle was to be the external examiner in mathematics for the University of Ceylon in the year that I sat my final degree examination. It amused me later on to think that Fred Hoyle would have read my examination scripts long before he ever set eyes on me.

Besides Eliezer, two other lecturers were a powerful influence on my mathematical and scientific development. One was an archetypal polymath — Douglas Amarasekara, a brilliant mathematician, a Cambridge Wrangler, whose interests spanned astronomy, art, music and literature. He considered himself to be a modern Socratic figure and often invited a small group of us to his house to listen to music, discuss books and paintings and to engage in Socratic type dialogues. Our regular group included myself, Asoka Mendis, who much later became an astrophysicist and Professor at UCSD, and H. Codipilly who became a researcher and Senior Economist working for the World Bank. All of us were in the Mathematics Honours class to which Amarasekara lectured. His lectures were unfortunately less inspiring than his Socratic sessions turned out to be. Another lecturer who inspired me with his style and elegance of presentation was P. W. Epasinghe who later became Professor of Mathematics at Colombo University, and is now the Science Advisor to President Mahinda Rajapaksa.

In the summer of 1960 I graduated with First Class Honours in Mathematics. This was the year in which the Commonwealth Scholarship Commission was set up in London with the aim of enabling the ablest graduates from the Commonwealth to come to British Universities to obtain their post graduate degrees. I applied for a Commonwealth Scholarship to do postgraduate studies in astronomy at Cambridge University. At the same time I also applied for a place to do a PhD in Theoretical Astronomy at Trinity College, my father's old College, and I was delighted to be accepted, and more so to be told that I would be supervised by Professor Fred Hoyle of

St John's College who was then Plumian Professor of Astronomy and Experimental Philosophy at the University of Cambridge. Whilst at the University of Ceylon I had already read two classic books by Hoyle: *Nature of the Universe* and *The Frontiers of Astronomy*, both of which had made an indelible impression on me. So when I received a handwritten letter from Fred Hoyle at my home in Colombo, recommending a list of books to read prior to coming to Cambridge in October 1960, I was naturally overjoyed.

CHAPTER 5

DESTINATION CAMBRIDGE

In September 1960 I found myself preparing to sail away from my native country for the second time in my life. But on this occasion I was facing the daunting prospect of venturing out into an unknown world, and into a foreign country all alone leaving my family and friends behind. The very thought made me homesick before I even started. I entered this new phase in my life with courage only because of the anticipation of intellectual reward in the end. This was a reward I dreamt of from a very early stage in my life — to follow in the footsteps of my father, and seek to understand the universe of stars and galaxies, studying astronomy at the best University in the world!

In 1960 the normal way (and the cheapest) to travel from Ceylon to England was by ship. Although air travel was rapidly coming into vogue it remained a considerably more expensive option reserved mostly for business travellers and the rich. On a warm September evening I sailed away from the port of Colombo aboard the P&O Liner *SS Orcades*, wistfully watching a palm-fringed coastline recede slowly into the distance. A two week voyage took us through the Suez Canal, via Naples, Gibraltar and Marseilles to Southampton. Nowadays such a voyage would be regarded as a luxury cruise. But my enjoyment of this new experience was hindered by homesickness as well as sickness due to rough seas.

I remember still the sense of trepidation that crept upon me as I stepped off the gangway to set foot on the soil of Southampton. From there I took the Boat Train to London and thence another train to my destination Cambridge. I arrived at Cambridge railway station on a brilliant autumn morning and took a taxi to the Porter's Lodge of Trinity College. I clearly recall the sense of awe that swept over me as I walked across the Great

Court of Trinity surrounded by the ghosts of the past, legendary names in both science and literature. This was the College of Isaac Newton, Bertrand Russell and William Wordsworth, and a brochure I had in my hand directed my attention to Newton's rooms above the porter's lodge, where his classic experiments on prisms and light were conducted.

When this initial sense of awe had worn out, the harsh reality of my new situation began to dawn on me. I found myself in the relatively comfortable Trinity lodgings for graduate students in Burrell's Field and for the first time ever I had to fend for myself. Apart from having all my fees taken care of, the Commonwealth Scholarship gave a princely £50 per month for living expenses. This would in modern money be worth at least five times more, so it was in fact a princely sum. In addition there was an allowance for buying books. I had to open a bank account, equip a kitchen with kettle and crockery and get provisions for breakfast, which was the only meal I fixed for myself. For all other meals I went to Hall, which at Trinity was as austere as it was spacious and cold. Portraits of illustrious figures from history adorned the walls, and meals were served on rows of long tables. After I established myself in College my first formal appointment was to see my tutor Dr. Robson who was in charge of graduate students.

Dr. Robson's welcome was eminently professional. He made me feel relaxed with a glass of sherry and offered a personal touch to our meeting by talking about my father's distinguished record as a student in 1930–1933. This, I later discovered, when I myself became a Fellow and Tutor at Jesus College, was a trick that had to be learnt! After my first lunch in the College Hall and a few other meetings with Junior Bursars and the like, I returned to my lodgings. Among a pile of inconsequential letters that awaited me was a handwritten letter by R.A. Lyttleton (Ray Lyttleton) of St John's College asking me to see him in his rooms to discuss matters relating to my supervision. That surprised me somewhat because I had already received a letter from Fred Hoyle in Colombo saying that he would be supervising me.

When I met Lyttleton at the appointed hour I learnt that Fred Hoyle was in the USA for most of the Michealmas term and that he, Lyttleton, would supervise me to start with. My encounter with Lyttleton was little short of formidable as I now remember it. After inquiring whether I had any specific problems in mind that I wished to work on, he informed me that astronomy was an extremely difficult subject to research. Most of the easy problems had already been solved, and what remained unsolved was insurmountably difficult! Only much later did I come to realise that these remarks were probably true to an extent if one confined one's attention to areas of classical astronomy that excluded applications of physics, astrophysics as it is now called. At this first meeting Lyttleton asked me to take a look at a rather abstract problem in the theory of stellar structure. He also very generously gave me a copy of his book on comets in which he described the "dust bag" theory of comets (which I still possess), following from work he had done in the 1940's. The book was in fact an elaboration of a paper he co-authored with Fred Hoyle on the accretion of interstellar dust by the sun. Lyttleton, it should be noted, was the person who turned-on Fred's interest in Astronomy at a time when he was beginning to find Nuclear Physics, the subject he first began working on, a fallow field of research in the 1940's. It should be recorded at this point that Fred Hoyle's collaborations with Lyttleton on the subject of accretion of gas from stars had a crucial historical relevance to my story. For the first time Fred became aware that the formation of molecules from atoms in the interstellar medium was an inevitable process. Hoyle and Lyttleton wrote a paper describing a mechanism by which large molecules would form but this was not accepted by the scientific establishment. Hoyle later told me that through the 1940's and early 1950's he had it deep within him that interstellar clouds were thick with organic molecules. But because the radio astronomers of the day maintained so adamantly that this could not be so that he abandoned the idea as a scientific project but turned it into the novel *The Black Cloud*. In this novel a dense cloud of interstellar dust and organic molecules endowed with intelligence approaches the sun and threatens to cause havoc with the climate and to endanger our very existence. Similar ideas based on science came to be pursued after I started the long collaboration with Hoyle which was now about to begin.

I now had enough reading to get on with so I decided to bide my time until Fred Hoyle returned from the States. I also attended a few advanced lecture courses that were being offered to students of the Mathematical Tripos, including Paul Dirac's course on Quantum Mechanics. All this was in preparation for when Fred Hoyle would return from the USA.

The next significant event I remember from my early Cambridge days was an invitation to tea by Jayant Narlikar, who later was to become a family friend. This was my first meeting with another of Fred's students who started research in the academic year 1960/61. When I met him in October, Jayant had already started serious work on a research problem in cosmology. There were two competing models of the Universe. One was the Steady State Theory and the other the Big Bang Theory. The former theory asserts that the overall average properties of the Universe remains the same at all times, so the expansion of the Universe must be compensated by the creation of new matter. The Big Bang Theory says the Universe originated in a gigantic explosion (a Big Bang) at a definite moment, a time that is now estimated as being 13.7 billion years ago.

My conversations with Narlikar made it clear that the 1960's were turning out to be a watershed era for cosmology. A right royal battle between the Steady State Theory of Hoyle, Bondi and Gold and the rival Big Bang Theory was set to begin. Cambridge Radio astronomers led by Martin Ryle had been claiming in the late 1950's that the Steady State Theory could be disproved by their study of the distribution of distant galaxies in the universe emitting radio waves (radio source counts). Hoyle argued that these measurements were far from secure, and indeed were most likely to have been contrived. It turned out that Ryle's early analysis of the matter was based on limited surveys and consequently poor statistics.

Throughout successive surveys of radio sources made at the Mullard Radio Observatory in Cambridge, Ryle and his team pressed home the point that there was enough evidence to disprove Steady State Cosmology. In a simplistic interpretation, the universe appeared, on the basis of radio source counts, to be more compact at the earliest epochs, favouring the concept of a Big Bang origin. Jayant Narlikar came on the scene amid gathering storm clouds, and the atmosphere in Cambridge was highly charged when I first arrived.

The insatiable appetite for denigrating a Steady State Cosmology of any type, often with the flimsiest of evidence, that I witnessed in the 1960's still puzzles me. I cannot help thinking that the reasons have a deeply cultural basis. The steady state theory effectively challenged the first page of Genesis, and this would naturally be strongly resisted by a predominantly Judeo-Christian culture. Without going into the technical details on either side of the argument my own cultural predilection was for a steady state universe of some kind. Such a cosmology is consistent with the philosophical world view that pervades the Indian subcontinent, and in particular it is in harmony with Buddhist traditions that are prevalent in Ceylon. Another aspect of the Cambridge debacle that astonished me was the component of personal jealousy that entered a scientific controversy. In my naivety I had believed that science was pursued in a cold detached manner, independent of personalities or social constraints. This was far from the truth. It was clear from the situation I witnessed that the two contesters in this argument were worlds apart in their personalities and ideologies. Fred Hoyle was a forthright and candid Yorkshireman; Martin Ryle a stiff upper lip product of the public school system. Never would their differences be reconciled.

When my turn finally came to meet Fred Hoyle, cosmological problems of enormous gravity would have been on his mind. As I walked from my lodgings on a frosty afternoon in December, across the backs of the river Cam, over a desolate playing field of St John's, I wondered what my long-awaited meeting would be like and what it might eventually lead to. Would I encounter a man weighed down by the strain of a huge controversy, preoccupied and terse; or a sparkling communicator, the author of the some of the most stimulating books I had read? But my anxieties were quickly dispelled when I arrived at the door of 1 Clarkson Close.

Fred Hoyle's wife Barbara greeted me with a warmth of affection and generosity that I could never forget. Fred, I was told, was just finishing an interview with a journalist and I was ushered in to their dining room to take tea with Barbara and her mother, Mrs Clarke. The ladies were so welcoming that in minutes I was completely at ease, and within half an hour, I felt that I had known them for a lifetime.

When the journalist had left I was taken into a spacious, thickly carpeted open-plan living room that also served as both library and study. Large patio windows extended over the entire length of the room on the far side, looking out on a slightly undulating lawn. There was Fred seated on his easy chair by the window, with a writing pad on his knee, fountain pen in hand scribbling a calculation with intense concentration. It was hard to decide whether he was pleased or not to be disturbed, but as Barbara introduced me as his new Ceylonese student, he switched into a more relaxed mode.

I cannot remember exactly how our conversation went, but I do recall a remark that embarrassed me such as "I hear you write poetry?" He must have seen such a declaration on the application forms that were sent to the

University. On confessing that I had published a slim volume of poems and had some poems included in an *Anthology of Commonwealth Poetry* that was published by Heinemann that year, he was noticeably impressed. We had established a connection at a deep level — an abiding passion for creative writing and a love of the English language.

We talked about poetry and politics, cricket, Ceylon and my professor in Colombo (C.J. Eliezer), who had been a Cambridge contemporary of Fred's. We even talked about the weather! In fact we talked about everything except science. I took all this to mean that he had not yet given much thought to what I might do as a research project, probably because of his preoccupation with other more incumbent matters at the time. He did however direct me to his own monograph on Solar Physics and a book by Cowling on Magnetohydrodynamics, but without any explicit application to think about. I left Clarkson Close that evening feeling greatly relieved as well as excited. And I had finally met one of my childhood heroes in science.

Within weeks of my first meeting with Fred I find myself immersed in problems of solar physics and soon write my first paper with Fred Hoyle "On the reversal of the Sun's polar field". The sun is essentially a giant magnet that with north and south poles that flip once every eleven years (the last reversal took place in 2013). In our paper, for which I made the calculations, we describe the mechanism by which this reversal of polarity occurs. Although this was an irrelevant side stream in the main flow of my scientific work that followed, the strength of this research led me to my first conference. Fred Hoyle and several other scientific luminaries were to lecture here, so it was for me, at the start of my career, a stimulating experience. The conference (which was a summer school) was organised by the Enrico Fermi Institute and the Italian Physical Society. Most memorably for me it was held at an exquisitely beautiful location — Villa Monastero in Varenna on Lake Como. Here I met several great and distinguished scientists, talked to them informally, and realised for the first time that famous scientists were ordinary people like you and me!. Among those I met was Fritz Houtermans who as early as 1929 had made the first calculation of stellar thermonuclear reactions that led Hans Bethe in 1939 (the year of my birth) to put forward his theory of nuclear energy generation in stars. Both the exquisite location of the conference and meeting people like Houtermans made the experience unforgettable to this day.

CHAPTER 6

A PROJECT TAKES SHAPE

In 1961 historic events were taking place in space exploration, mainly on the Russian side. The Soviet cosmonaut Yuri Gagarin becomes the first person to orbit the Earth aboard the spacecraft *Vostok 1*. The long series of *Venera* probes to the planet Venus were launched, the results from which I would later use to infer the possible presence of microbial life in the Venusian atmosphere.

I was beginning to enjoy my life as a research student in Cambridge reading widely and attending a variety of advanced and specialised lecture courses. I enjoyed the new freedom of being able to learn without the overhanging threat of examinations. This was also a time of readjustment to a new country, a new culture and a different way of life.

The way of life in the Western world also seemed to be undergoing rapid change. There was an expansion of global air travel and the advent of the concept of multinational industries. The stage was also being set for the modern world of instant communication, although the internet itself was still a couple of decades in the future. More importantly, in relation to my future research, we were on the threshold of the Space Age that was destined to transform astronomy forever.

A less tangible transformation of social attitudes had also begun. The election of John F. Kennedy as President of the United States was seen as a new beginning. A quiet confidence and optimism began to pervade Western society. The lean post war years of the late 1940's and 1950's that I experienced firsthand as a child living in London, had given way to an unprecedented economic boom. This had an effect on all our lives and a feel good factor was evident in people's psychology.

Problems were, however, looming large on a distant horizon — the escalation of the Cold War, the Arms Race, the growing dissension in South Africa, conflicts in the Middle East and the slow beginnings of global terrorism. But all this made no impression yet on the day-to-day lives of people in the West. Along with slogans for combating communism came clamours for freedom, liberalism and permissiveness. *Lady Chatterley's Lover* found its acrimonious entry into the literary canon, and the cause of feminism was taken a step further with the election of the World's first woman Prime Minister in Ceylon, Mrs Bandaranaike, following the assassination of her husband S. W. R. D. Bandaranaike. For me the new found sexual freedom that was given abundant expression on the streets of Cambridge came as a shock to my sequestered upbringing in an ex-colony with distinctly Victorian values.

The 1960's could be seen as the halcyon days for Science. Most areas of research advanced steadily and with the same confidence that characterised society as a whole. The dawn of the new decade was marked by several notable technical triumphs. The laser, a light source of unprecedented intensity, made its debut in 1960. The quark theory proposed that particles like protons and neutrons, hitherto thought to be the fundamental units of matter, were in turn made up of even more basic units called quarks. The Higgs boson, the particle responsible for giving mass to all other particles, was proposed. This particle was a theoretical concept until 2012 when the Large Hadron Collider near Geneva actually detected it. For this discovery Sir Peter Higgs and Francois Englart were awarded the Nobel Prize for Physics in 2013.

Exploration of the Moon was also progressing briskly at the time with a healthy competition between the Americans and the Russians giving it a stimulus. Neil Armstrong and Buzz Aldrin walked on the Moon. The Space Age was well under way.

All these developments seemed to flow naturally and with ease, but it should not be forgotten that the groundwork — perhaps the hardest groundwork — in many areas was already laid earlier in the 1950's. It was in the mid-1950's that radioastronomy began, and amongst its rich harvest are discoveries that would turn out to be germane to the questions that will be addressed later in this book. For instance the study of how hydrogen gas is distributed in the galaxy, and the discovery of molecules connected with life, all stemmed from developments in radioastronomy. As I mentioned earlier the mid-1950's also saw the successful completion of the work by Geoff Burbidge, Margaret Burbidge, Willy Fowler and Fred Hoyle that led to an understanding of how the chemical elements, including the elements of life, are formed from hydrogen in the deep interiors of stars. At about the same time, there was the monumental discovery by James Watson and Francis Crick of the famous double-helix structure of our genetic material — DNA. Shortly afterwards Fredrick Sanger analysed the nature of a protein (insulin to be specific), showing its detailed sequence of constituent amino acids; and Harold Urey and Stanley Miller completed their classic experiments in which they showed how the most basic chemical building blocks of life might be synthesised from inorganic matter.

It might seem ironic that the burgeoning of liberal traditions in society had little effect in promoting a spirit of free and unfettered inquiry on the scientific scene. Although science in general and Astronomy in particular, continued to flourish, there was a decline in the development of new ideas. Newly invented experimental techniques generated a vast body of factual data. But there was precious little being done in the way of a critical re-appraisal of old hypotheses. The belief grew that the really important problems were either close to being solved, or that they were really so hard that we should not even begin to worry about them. In practical terms it was a case of turning out more of the same type of factual data, attempting to consolidate existing theories, seeking to tie up the last loose ends. An openness of mind required for a vibrant scientific culture was noticeably lacking.

Such then was the general backdrop against which my specific experience of research continued through the early part of 1961. As mentioned earlier Fred Hoyle was engrossed at the time with trying to resolve the radio source count conflict with Martin Ryle — a battle that was becoming ever more bitter. Fred Hoyle, however, never slackened his determination to defend steady-state cosmology, in the conviction that the alternatives were far less credible, and the data adduced in their favour either contrived or indeterminate.

Although Fred Hoyle saw Cosmology as his prime battle field in the 1960's, his interests were by no means restricted to Cosmology. His knowledge and experience across the entire field of astronomy and astrophysics was impressively encyclopaedic. By 1961 there were few areas of this subject that had not been embellished by his imaginative genius in some way.

By working on a wide range of problems Fred Hoyle's attention may have been diverted from the bitterness of the cosmological conflict. Another welcome diversion he had was an alternative career as a science-fiction writer. In 1959, a year before my arrival in Cambridge, he had already published his novel *The Black Cloud*, which in a sense was a fore-runner to his much later collaborations with me on life in the cosmos. I acquired a copy of *The Black Cloud* in the spring of 1961 and remember reading it avidly, particularly because I was able to connect the political nuances and intrigues described in the book with what I had seen enveloping Fred's life in recent months. The hard scientific content of the novel also impressed me enormously. His arguments for complex molecules acting collectively as intelligent entities intrigued me even before I began to work on organic molecules in space. Fred Hoyle's science fiction novels were not merely plausible they were entirely in the realm of what was possible according to what was definitely known (or *almost* definitely known) about the real world.

In the spring of 1961 when I was still groping to find a definite research problem to tackle, thoughts about the nature of interstellar matter began to enter my head. I saw Fred Hoyle a couple of times and recall expressing more than a passing interest in the scientific content of *The Black Cloud*. He informed me then that he had found it difficult to publish his calculations that showed the hydrogen molecule and other molecules to be abundant in interstellar space, and that his novel was a way of expressing these ideas at a time when science would not countenance them. This is unfortunately the way of contemporary science: the rein of authority imposed by a scientific orthodoxy stamps out any attempts to challenge it. Such an attitude is often to the detriment of progress, as we shall see.

I discovered that Fred Hoyle was at this time also engaged in another science fiction project with TV producer John Eliot. Together they were writing the script for *A for Andromeda* which became a highly successful BBC TV serial when it was broadcast in 1961 and 1962. Later it was to become a cult classic of sci-fi fanatics. In the script, a newly built radio-telescope picks up intelligent signals from the constellation of Andromeda which are interpreted as instructions for building a gigantic super-computer.

Once built, the computer begins to relay the information it receives from Andromeda and the security of humankind falls under dangerous threat. By presuming the presence of intelligent life in a far-flung corner of the Universe, albeit through the guise of science fiction, Fred Hoyle was moving unmistakably towards some form of astrobiology as early as 1961.

When my interest in *The Black Cloud* matters became evident, Fred directed me to a review article on "Interstellar Matter" by Jesse Greenstein, a distinguished astronomer who was at the time working at the Mt Wilson and Palomar Observatories in the US. Greenstein had discussed the composition of interstellar dust, considering the arguments for the two separate classes of dust grain — icy particles and metallic iron particles — as I have previously mentioned. At this time, however, there was no clear idea as to how either of these dust grain types could populate interstellar clouds. My reading introduced me to many questions that seemed to be in urgent need of answering.

In my initial discussions with Hoyle at his home in Clarkson Close it was clear that he was not happy with any of the theories that were around at the time. My 40-year long scientific collaboration with Fred Hoyle was now about to properly begin. I arrive at the door of 1 Clarkson Close on a sunny Spring evening in 1961 and, as usual, I am welcomed by his wife Barbara Hoyle. Before taking me to see the great man, I was, as usual, brought up-to-date with the family's doings. Barbara seemed more socially disarming than usual when she announced that *she* had decided that I should accompany Fred on a walking holiday in the Lake District. Here I could experience first-hand the romance of the Wordsworth country I had read so much about, and, as she put it "you men could plan your work". A destiny of friendship would seem to have been pre-ordained.

I was naturally excited and looked forward to the prospect of spending time with Fred in the Hills. I find it hard to decide between hills and sea as locations I would choose to commune with Nature. From an early age I acquired a yearning and love for both. I was born and grew up a mile or so from the Indian Ocean coastline. But my most memorable holidays were in the hill country of central Ceylon with mountain slopes clothed with lush green plantations of tea. I regard the sea and the hills to reflect life in their different ways. The rough oceans reflect the turbulence of our lives, whilst the calmer seas depict our quieter moods of reconciliation with the world. The hills, on the other hand, are a symbol of tranquility and a retreat from the trouble of the world:

> "For, oft when on my couch I lie
> In vacant or in pensive mood,
> They flash upon that inward eye
> Which is the bliss of solititude."

— William Wordsworth

The mountains are also the closest to being symbols of eternity within our reach, formed and structured over aeons of geological time.

My recollections of the most brilliant night skies visible in the hill country still haunt me. Nowhere else except perhaps in the midst of the American desert or on the top of Mt Palomar had I witnessed a comparable sight.

At the end of March 1961 Fred Hoyle drove me in his two seater sports car that he had at the time. We arrived at the Old Dungeon Ghyll Hotel, then a modest guest house in Great Langdale (near Ambleside), a building that was over a hundred years old. It has since been transformed into a fine hotel, but in those days it was a serious hikers' retreat, run by the mountaineer Sid Cross and his wife. It was early evening and there was just enough time to unpack before we sat down to a sumptuous dinner cooked by Mrs Cross. The evening meal at the Old Dungeon Ghyll was the focal point of the day and was more than ample compensation for the somewhat Spartan accommodation that it offered.

Visits to the Lake District and walking on the hills was Fred Hoyle's way of escaping from the tribulations of academic politics. In the mountains of the Lake District he found repose and time to think. Here he would experience for a while what he saw to be the real challenges of life, those that a mountaineer had to face, struggles with the elements and the terrain. In this way the irrelevant squabbles in the cloisters of university would fade momentarily into insignificance.

On this particular trip Fred appeared to have a well-defined plan which included me and my project for a PhD. After dinner on our first evening the conversation in front of a blazing log fire turned briefly to the scientific matters we had discussed at Clarkson Close. What would be an acceptable alternative to ice and iron for the composition of cosmic dust? Mesmerised by the flames rising above the burning and sputtering logs, I remember posing a question without too much thought. Could the dust in space be carbon, like the soot that lofted up from the fire into the chimney? Fred's eyes appeared to light up, with excitement, or disbelief, I could not say. Carbon in the cosmos had a particular resonance for Fred. It was his prediction of a resonance (at energy 7.65MeV) in the nucleus of carbon that opened up new vistas of astronomical thought. Fred, together with his collaborators Willy Fowler and Geoff and Margaret Burbidge told the world how carbon was in fact synthesised in nuclear reactions taking place in stars, and expelled into interstellar space by exploding supernovae.

But carbon in interstellar dust, as I had flippantly suggested, did not make an immediate impression on Fred, no more than iron did. We all had become used to a mode of thinking, following the trend set by Dutch astronomers of the 1940's that the bulk of interstellar dust had to be formed *in situ* in the gigantic diffuse gas clouds of space. And of course oxygen is more abundant there than carbon, so most of the carbon of interstellar space would tend to be tied up in the strongly bound molecule carbon monoxide (CO). I would guess that these were the thoughts that raced through Fred's head at my seemingly outrageous suggestion of a possible carbon composition of dust.

The following morning after breakfast I joined Fred on his habitual trek over the hills, heading I was told to Bowfell. He had decided that the more ambitious climb to Scafel Pike was not for first time hikers like myself. Although I was armed with a pair of mountaineering boots and an appropriately warm anorak and headgear, I felt hopelessly ill-equipped for the task ahead. Fred had a battered-looking rucksack on his back in which he carried sandwiches and bars of chocolate supplied by Mrs Cross. I remember Fred saying that a single bar of chocolate would have enough calories to make good the energy we would expend all the way up to Bowfell Pike and back.

We set out under an azure blue sky almost cloudless save for a stray cumulous cloud or two drifting lazily overhead. The walk began as a more or less gentle amble along the valley which suited me fine, but before long the proper climb began and I was finding it increasingly difficult to keep up with a seasoned hiker like Fred. I believe that Fred noticed my predicament and accordingly modified the route to take us only so far as a lesser peak than Bowfell on this first day.

The gentler walk was welcome, if only to enjoy my first experience of the exhilarating scenery of the Lake District. The landscapes that I had dimly apprehended, thousands of miles away in Ceylon, through the works of the romantic poets, now revealed themselves in all their sheer glory. Walking here in the company of Fred Hoyle had a particular poignancy. I recalled Wordsworth's Preface to his "Lyrical Ballads" in which he compares the aspirations of the poet and the scientist:

> "The knowledge of both the Poet and the Man of Science is pleasure; but the knowledge of the one cleaves to us as a necessary part of our existence, our natural and unalienable inheritance; the other is a personal and individual acquisition, slow to come to us, and by no habitual or direct sympathy connecting us with our fellow beings... ...Poetry is the first and last of all knowledge — it is immortal as the heart of man... ."

Quite early in my association with Fred Hoyle, I could see how the poet and scientist could come together in a single individual, and I hoped myself that I might be such a composite person.

Some three hours later, we stopped for a spot of lunch. As every mountaineer will tell you the Lakeland weather is extremely fickle, changing suddenly without any warning. So it seemed to be today. The sun had slipped behind an ominously dark cloud and a shower of rain might have been imminent. We found a flat crag to sit upon and as Fred unpacked our sandwiches from his rucksack he glanced thoughtfully at the grey sky. When I casually enquired, "Is it going to rain?" I could not have predicted that his answer would turn out to be a defining moment in my scientific career.

"Not necessarily", said Fred, "These clouds could be saturated with water vapour, but for rain to fall, condensation nuclei are required. These could be charged molecular fragments (ions) or fine dust, but such condensation nuclei need to be present before rain could form". With a moment's further reflection he added, "Some people have argued that meteor dust could supply nuclei for rain".

From my long experience of Fred Hoyle I have since realised that a thought such as this rarely remained in isolation in Fred's head: he would begin to make connections with the widest range of problems. With a little prompting from me the connection with dust in interstellar clouds soon came to the fore. If it is difficult to form water droplets in the densities that prevail in the terrestrial atmosphere, how could ice particles condense in the exceedingly tenuous clouds of interstellar space, where hydrogen densities are in the range 10–100 atoms per cubic centimetre? Was the nucleation problem really solved for the case of ice grains in interstellar space? These were questions we pursued in the remaining few evenings by the fireside in the lounge of the Old Dungeon Ghyll Hotel. Fred's writing pad and pen were used for extensive scribbling and between us we decided that the problem could not really have been solved by the Dutch astronomers. It seemed incumbent on us, therefore, to find a denser place than the interstellar medium to resolve the problem of how interstellar dust grains are formed. And this was the beginning of a long trail.

CHAPTER 7

GRINDING OUT THE DETAILS

After returning from my visit with Fred Hoyle to the Lake District I spent most of the next year (1961–1962) grinding out the details of a revolutionary new model of interstellar dust. I discovered fatal flaws in the then fashionable theory that interstellar dust was comprised of ice grains. But a great deal of hard work was needed to convince our critics that this was so. This is typical of the way that scientific advances are made. A flash of insight or intuition may point the way to new possibilities but the hard slog of working out the details must follow before the final product can be delivered.

Although by now I had become accustomed to my exile in a foreign land, I was never really able enjoy the physical climate — inclement weather and persistent grey skies, that classic ice-breaker of English conversation. The long dark evenings of winter still, after more than 50 years, make me yearn for a more equable climate and to the beaches or hills of Sri Lanka. Nevertheless, I had no time to really feel homesick, being always deeply involved in the work in progress. And there was also a great deal of background astronomy I had to learn.

Stars emit visible light as well as radiation invisible to the eye — from X rays and ultraviolet to infrared, microwave and even radio wavelengths. The distribution of the energy radiated at these various wavelengths depends on the temperature of the radiating objects. Our sun, which is at temperature of about 6000°C at its surface, emits most of its radiation in the visible part of the spectrum. This is the circumstance that gives the chance for life to exist on the Earth. The main absorption wavelength of chlorophyll, the green colouring substance in plants, is close to the peak emission wavelength of the sun. The sun's energy is thus efficiently taken up by plants to convert water and carbon dioxide into carbohydrates, and this is at the base of the food chain for all life on the Earth. Other stars, if they are cooler, emit most of their energy in the far red or infrared wavelength range; if they are much hotter they would emit mostly in the ultraviolet. Superposed on this continuum of radiation emitted by these stars is a host of very sharp and narrow absorption or emission lines, giving astronomers all the information necessary to deduce the types of chemical elements of which stars are made.

Starlight travels through vast distances, hundreds of light years, to reach us. On the way this light must cross one or more interstellar dust clouds. If the clouds happen to be too thick in dust they may effectively block out all the visible light from the stars. This is indeed the reason for the dark appearance of clouds of various shapes — elephant's trunks, horse's heads or eagle's nests — seen against the background of stars in the Milky Way. For clouds that are not so dense, the light from stars traversing through them is partially absorbed and scattered by dust particles in ways that leave signatures of their properties on the light that is received. The dim light of a street lamp seen through a fog has imprinted on it tell-tale signs that reveal the chemical properties and sizes of the intervening dust particles and molecules in the air. In the same way the dust in interstellar space leaves its imprint on the transmitted light from stars.

Deciphering the nature of cosmic dust requires (1) astronomical observations of stars over a wide range of wavelengths, (2) modelling the possible or plausible compositions of the dust, and (3) performing calculations of the scattering and absorption properties of dust to discriminate between various models. This is precisely the type of research that was to occupy a large part of my professional life as an astronomer over the next five decades, and lead eventually to a new theory of the origins of life. The "Black Cloud" of Fred's novel was to enter the realms of hard science.

The dark clouds of deep space had, however, to be set in the context of what else there existed in the Universe. The Milky Way itself, against which these clouds can be seen, our Galaxy, is a few hundred thousand light year-wide collection of hundreds of billions of stars, each one more or less similar to the sun. And our galaxy is one of many billions of similar galaxies that populate the observable Universe that is unimaginably vast.

Interstellar dust clouds are by no means restricted to our own galaxy — our Milky Way system. External galaxies often show conspicuous dust lanes. We know that interstellar clouds in our galaxy as well as in external galaxies are stellar nurseries, the birthplace of new stars and planets. Therefore they must clearly have a very special importance in the scheme of things — an importance that was not fully recognised in the 1960's.

An interstellar cloud is on the average about ten light years across in size, and the typical separation between neighbouring clouds is about 300 light years. There is a fairly wide spread in the sizes of these clouds and in their structure: some are more compact and uniform in their disposition, whilst others are extended and irregular. The more extended clouds appear as giant complexes, showing a great deal of fine-scale structure as cloudlets and filaments. These are the so-called "giant molecular clouds" of which the molecular complex in the Orion nebula is an example.

An interstellar cloud may contain anywhere from ten to many millions of individual atoms per cubic centimetre. Even the higher values in this density range are considerably lower than the densities that can be attained in laboratory vacuum systems. So it should be remembered that all our intuitive ideas of how gases behave under normal conditions may prove wide off the mark when it comes to understanding what happens under the extremely rarefied conditions of space.

Hydrogen makes up the overwhelming bulk of material in interstellar clouds and occurs in one of three forms: neutral atomic hydrogen (intact atoms with no electrons lost), ionized hydrogen (atoms stripped of their outer electrons), and molecular hydrogen (atoms paired in molecular form as H_2). Molecular hydrogen was first detected using ultraviolet spectroscopy in the late 1960's after my own researches into interstellar matter had begun, although its existence was predicted by Fred Hoyle in the 1950's and indeed exploited in *The Black Cloud*. Hydrogen molecules are to be found mostly in the denser interstellar clouds that are able to screen off the ultraviolet starlight that would otherwise destroy them. A large fraction of all the hydrogen in the galaxy is found to be in molecular form, H_2, and the total mass of the molecular hydrogen is billions of times the mass of the sun.

But what else is there in interstellar clouds besides hydrogen? Information derived from several sources, including studies of the spectra of the sun and stars, and the direct examination of meteorites (rocks of extraterrestrial origin) all have a bearing on unravelling the overall composition of interstellar material. Next to hydrogen, in order of abundances, comes the element helium, which accounts for close to a quarter of the total mass of interstellar matter. But from our point of view this element is inert, chemically unreactive and therefore uninteresting. Then came the group of chemical elements carbon, nitrogen and oxygen that together make up several percent of the mass of all the interstellar matter. It is these elements that are of course crucial for life. Indeed life depends for its function on the unique range of properties of the carbon atom including its high levels of chemical reactivity and its ability to combine into many millions of interesting carbon-based compounds. Next in line are the elements magnesium, silicon, iron and aluminium, which again account for a percent or so of the total interstellar mass. Then a group including calcium, sodium, potassium, phosphorus is followed by a host of other less abundant atomic species. All these chemical elements are synthesised from hydrogen in the deep interiors of stars in the manner worked out by Fred Hoyle and his colleagues in the 1950's. The synthesised elements are injected into interstellar space through a variety of processes, including mass flows from the surfaces of stars. In the case of the most massive stars, the end product of their evolution is a supernova (an exploding star), and it is through supernova explosions that life-forming chemical elements are mostly injected into the interstellar clouds.

The discovery of interstellar molecules (assemblages of atoms) by methods of radioastronomy, millimetre-wave astronomy and infrared astronomy got properly under way long after my own researches into these matters had begun. Next to molecular hydrogen the second most abundant and widespread molecule in space turns out to be carbon monoxide. Today we know that a vast array of complex organic molecules exist in huge quantity, and these discoveries as they progressed from the 1970's to the present day provided the thrust of the theoretical work that I became involved in with the passage of time. In the denser interstellar clouds, particularly in clouds associated with newborn stars, vast amounts of water in gaseous form are found. Water is an important molecule for life, and its close association, along with the organics, with newly formed stars and planetary systems would have a vital relevance to my story.

The spatial distribution of interstellar molecules in the galaxy shows wide variations depending on physical conditions such as ambient temperature and density as well as the proximity of clouds to hot stars. As a rule, denser and cooler clouds contain the larger and more complex molecules, whereas lower density clouds and those nearer to hot stars have simpler molecular structures. A region that is particularly rich in organic molecules (molecules that could be connected with life) is the complex of dust clouds in the constellation of Sagittarius, located near the centre of the galaxy. It is in this region that the first detection of an interstellar amino acid, glycine (a component of proteins) was reported, as was the molecule of vinegar and a sugar glycolaldehyde. It is also here that, many years later, the signature of bacterial type material was discovered by my brother Dayal and David Allen deploying the Anglo-Australian telescope.

An important class of organic molecule that is found to be present in vast quantity in space are the so-called polyaromatic hydrocarbons (or PAH's). These molecules (just as CO) are well-known by-products of the combustion of fossil fuels as occurs for instance in automobile engines. They are essentially combustion products of living material and are largely responsible for the suffocating smog that pollutes our larger cities. We shall argue later that even in interstellar space such molecules are most likely to have a direct biological connotation, possibly representing the break-up or degradation products of biological material.

Comparatively high densities of organic molecules tend to be associated with regions of the galaxy where new stars (and presumably comets and planets) are forming at a rapid rate. The Orion nebula is a spectacular example of such a region where young stars are evident in large numbers, some even with discs of newly formed planetary material seen around them. Large quantities of organic molecules are associated with the denser parts of the Orion cloud complex, and it would be tempting to link the formation of such molecules with the formation of stars, planetary systems, and perhaps life itself as I shall elucidate later.

In addition to atoms and molecules, interstellar clouds contain an all-pervasive and enigmatic dust component to which I have already referred. Astronomy had struggled to understand the precise nature of this cosmic dust and to discover the circumstances under which such particles are formed. In the Autumn of 1961 when I started reading on these matters in earnest I discovered that it was almost an article of faith amongst astronomers that interstellar dust grains were comprised of dirty ice material — frozen water with perhaps a sprinkling of other ices — ammonia ice, methane ice and a trace quantity of metals. Furthermore, the firmly-held belief was that these particles had to condense, more or less continuously from the gaseous atoms and molecules that were present in the interstellar clouds.

The basic ground rules in this area of astronomy had been laid down in two classic papers and a PhD thesis written in the mid1940's by the Dutch astronomer H.C. van de Hulst to which I have referred earlier. My own researches had led me to a large body of work in an area of physics known as "homogeneous nucleation theory" that was immediately applicable to the interstellar problem. It was easy enough to check the claims of van de Hulst and his colleagues against straightforward predictions of the theory. I soon discovered that van de Hulst's ideas were deeply flawed in several important respects. I was quickly able to verify Fred Hoyle's conjecture about the nucleation problem — the insuperable difficulty of forming dust in the exceedingly tenuous gas clouds of interstellar space. The rain cloud analogy I was introduced to in the Lake District applied here, only more dramatically. When I communicated these findings to Fred Hoyle it was evident that he was pleased. His immediate response was that I should see van de Hulst in Leiden and confront him with all the technical details and the difficulties as we perceived them.

So in November 1961 I made the journey by boat from Harwich to the Hook of Holland, and thence to Leiden. I was much taken up by Leiden as a city which was not unlike Cambridge in many ways. It is an historic University town, somewhat younger than Cambridge — the University being founded in the 16th century — but with a very similar ambience. There were as many bicycles in Leiden as in Cambridge, but unlike Cambridge's single water way the River Cam, Leiden was criss-crossed with boat-lined canals. This city is the birthplace of Rembrandt and the city is saturated with art galleries and museums, which would take weeks of sightseeing to exhaust.

But the purpose of my visit was to tackle the great man H. C. van de Hulst. This too turned out to be memorable, albeit in a somewhat negative way. As affable as he was, it was clear at the outset that he was not inclined to engage in any sensible scientific dialogue with me. I sensed that he was possibly affronted by my challenge of his views. Only much later did I come to appreciate that Professors in Europe were traditionally placed on a high pedestal, their authority rarely questioned by younger members of staff let alone students. Be that as it may, van de Hulst in 1961 seemed unable or disinclined to offer a defence of his ideas of grain nucleation that he had pioneered a decade earlier. Perhaps he had lost interest in the problem, having moved on to new areas of research. He promptly directed me to his colleague, Professor J. Mayo Greenberg, at the Rensellear Polytechnic in Troy NY, who had now taken up the cudgels of defending the old ice grain theory against all its critics, and as we shall see later, against all the odds.

My visit to Leiden and conversations with van de Hulst gave me added courage to pursue alternative ideas on the nature and formation of interstellar dust. Faced with the almost insurmountable difficulty of seeing how ice grains could form in interstellar space I took a different approach. What if the dust was not made of water ice but of carbon, similar to the particles of soot that were rising into the chimney from the log fire we watched at the Dungeon Ghyll Hotel in the English Lake District? In this case the formation of carbon dust could occur at much higher temperatures, perhaps, for instance in the outer atmospheres or envelopes of some cool stars?

But what do the astronomical observations really tell us about the properties of interstellar dust? What are their precise optical characteristics? How do they behave in relation to the scattering and absorption of starlight? I have already noted that dust shows up as conspicuous patches of obscuration against the background of distant stars. But several more precise quantitative statements about the nature of the dust were already possible in the 1960s.

The earliest attempts to obtain measurements of interstellar dimming — or extinction, as it is called — of starlight were made in the 1930's. In the 1960's the most extensive modern data on interstellar extinction came from the meticulous observations of the Indian Astronomer (and my friend) Kashinath Nandy at the Royal Observatory in Edinburgh. It was found that at a single wavelength close to 4500Å (in the blue region of the spectrum) the dimming of starlight amounted to a reduction of intensity by a factor of about two for every 3000 light years of passage through interstellar space. From this one piece of information alone it was easy to infer that interstellar dimming could only be reasonably attributed to solid particles that have dimensions comparable to the wavelength of light, that is a few Ångstroms (1 Ångstrom = 10^{-8} cm.)

With the advent of new techniques in observational astronomy it became possible to measure accurately how interstellar dimming due to dust varies with the wavelength of light. This relationship between extinction and the wavelength of light — what astronomers call the extinction curve — provides an important item of information to tell us about the properties of interstellar dust grains. In 1961 the extinction curve was known only over a very limited range of wavelengths — from about 9000Å in the red end of the spectrum to about 3300Å in the near ultraviolet.

Over much of the visible wavelength range it was known that the opacity (light absorbing properties) of interstellar dust was proportional to the inverse of the wavelength — in other words, when the wavelength was doubled, the opacity was approximately halved. And most remarkably, it turns out that precisely the same type of relationship, the same extinction curve, was found to hold over wide areas of the sky. This means that we must have dust with almost identical sizes and properties throughout large volumes of galactic space.

The limited range of wavelengths for which the "extinction" data was available in 1961 (due largely to the work of Nandy) allowed for a wide range of possible dust models, including ice (of the type proposed by van de Hulst) and iron (of the kind suggested by C. Schalen). For each of these models, however, the sizes of the dust particles had to be fixed within a very narrow range. To match the astronomical extinction data to any particular model one needed to calculate the scattering and absorption cross-sections of particles of various radii using the Electromagnetic Theory. Such calculations that had already been made for iron grains and ice particles established these models as being *possible* candidates for interstellar dust, with ice particles having the edge over iron in certain crucial respects. Particularly so because it could be argued that there was insufficient iron in the galaxy to make up the required mass in the form of the interstellar dust.

In 1961–1962, in the days before personal computers, the calculation of the optical properties of any new grain model was a major computational undertaking, using formulae derived by the German physicist Gustav Mie in 1908. My task was to program these so called Mie formulae for use on a high speed electronic computer. This is what I set out to do in the winter of 1961 for a preliminary exploration of carbon dust models. I devised a computer programme in the then popular language FORTRAN to carry out the calculation, and ran this on the EDSAC2 computer of Cambridge University.

Computers in those days were large clumsy devices. Transistors and printed circuits had not yet been invented and the use of vast numbers of thermionic valves meant that computers were not only large and sprawling, they had also to be cooled efficiently with stacks of air-conditioning fans. This, together with the constant clattering of card readers and output devices, made the environment of the Computer Centre a very noisy place indeed. EDSAC2 occupied several large rooms in the Computer Centre and had less computing power than that of the average desktop computer, or even the smart phones of today. Although my problem was by no means a very difficult one, each computing run on EDSAC2 had to be booked in advance, so developing a new program, particularly debugging it was a tedious affair. After numerous trips to the computer centre, I made a remarkable discovery. I found that as long as the diameter of carbon particles were less than about a tenth of a micrometer, their predicted extinction behaviour was almost indistinguishable from the interstellar extinction observations as they were then known. It was also possible to calculate how much in the form of carbon grains we required, and the answer turned out to be one to two percent of the total mass of the interstellar clouds. This was consistent with the availability of carbon in interstellar space.

The next question to be asked is: where are these carbon grains produced? Searching for suitable cosmic locations, it seemed natural to turn to cool giant stars that are carbon rich. These would be the carbon stars, the so called R and N variable stars. The N-stars have surface temperatures that varied cyclically between 1800 and 2500 K over a period of about a year, and they are known to have more atmospheric carbon than oxygen. Thus although oxygen would link up with C to form CO, there would be an excess of C that is able to condense into solid particles when the temperature fell below some critical value. Again a computer programme had to be used to determine the physical state of the excess carbon as the temperature varied between 1800 and 2500K in the star's pulsation cycle. I showed that carbon particles were indeed able to nucleate and grow in the stellar atmosphere as soon as the temperature fell towards 2000K.

We further showed that particles of radii of a few hundred Ångstroms would grow and be expelled into interstellar space, the expulsion being caused by the pressure of light from the parent star. There was strong evidence from astronomy that pointed to the existence of dust around such carbon-rich stars. The variable and highly luminous carbon star R Corona Borealis is a spectacular example. Here we see direct evidence of a star erratically puffing out clouds of carbon soot into the interstellar medium, exactly as I saw the soot rise into the chimney from the coal fire at the Dungeon Ghyll Hotel.

In the summer of 1961 Fred Hoyle was to go away to California as he was accustomed to do, so I asked if I could spend a couple of months back home during this time. The Commonwealth Scholarship Commission that was sponsoring my studies did not usually encourage such home leave, and they felt that their scholars should be either working at their research through the vacations, or experiencing local hospitality of families that they were often able to arrange. My desire to visit my parents was so strong that Fred wrote the following letter to Mr. Foster:

"Dear Mr Foster:

I am writing in support of N C Wickramasinghe's application to return home for the long vacation. During the past three academic terms he has put in a great deal of hard work and his proposed visit to his parents could be a welcome break. As I, myself, will be away in the US for the whole of the summer there would be a break in his research programme in any case. In my opinion it would be better for him to return home than to be at a loose end in Cambridge.

Yours sincerely
Fred Hoyle"

In the event, I was granted leave of absence for three months, and was soon off on a plane bound for Colombo. The person seated next to me on the flight by a remarkable fluke turned out to be none other than Arthur C. Clarke, doyen of science fiction, futurist, and inventor of the idea of communication satellites. We soon got into conversation and began to find many common interests. Arthur Clarke knew Fred Hoyle. They shared a common publisher in W. Heinemann. He was also very interested in the ideas of carbon dust in the cosmos that I had just begun to explore. I recall a casual remark he made at the time that carbon dust must of course mean that there is life out there in space. I did not appreciate the importance of his remark at the time, but in his style of being a futurist he turned out unwittingly to be correct!

Arthur Clarke had lived in Sri Lanka from 1956 until his death in 2008. He followed the progress of my theories of cosmic life with great interest, and not surprisingly he was amongst my strongest supporters. In August 1962, when I was back in the UK, he invited me to visit him at his brother's house in London. The weekend I spent there remains immensely memorable. We talked at length about everything from the Second World War to cosmology, space travel and of course aliens. His views were often eccentric, but always original and entertaining. Many years later on my frequent visits to Sri Lanka, I would always go to see him and have long and interesting discussions on various subjects. I also worked with him in the 1980's when he was on the Board of Management of the Institute of Fundamental Studies of which I was the founder Director.

CHAPTER 8

LAUNCHING THE CARBON DUST THEORY

On every occasion when I have returned to Ceylon (Sri Lanka, now) after any lapse of time, I am overcome by a surge of emotion, recalling a period of my life that would always remain the happiest days of childhood. My first return from a chilly summer in Cambridge to sunshine and the warm embrace of family was particularly poignant, as I recalled thus:

"The sun beat down more harshly
Today, the day of my return.
The crisp hot pebbles of the pathway
Creaked under alien footsteps;
And the watchdogs barked savagely –
The scent of an old master long since
Wrung out from their handkerchief memories.

Even your arms smelt strangely of Time's passage –
The smell of fallow fields
Hung out in the mid-day sun.

The old folk, as always, sat on the verandas,
Puffing their native cigars,
Eschewing avidly the day's newspapers,
The men grown balder, the women greyer and more parched,
Some dead.

The sun saps away the last dregs of life.... ."

Visits back home during my student days were always fraught, particularly the inevitable farewells that had to eventually follow. After my summer sojourn in Colombo I returned to Cambridge refreshed and ready for more research at the start of the Michaelmas term 1962. Fred Hoyle had also returned from the US at this time, but throughout much of this term he was preoccupied with a mammoth project to investigate how the entire periodic table of chemical elements could be formed in stars. This did not, however, deter him from an active engagement in our joint project on carbon grains as this quite separate story began to unfold.

It did not take more than a few months after returning from the Lake District for us to write our joint paper entitled "On graphite particles as interstellar grains". This was submitted to the *Monthly Notices of the Royal Astronomical Society* and was selected for presentation at the meeting of the Society that was to be held at Dublin in April 1962. My formal presentation of the paper (the first in my career) was well-received, and with it the new theory of carbon grains was formally launched. The paper appeared in print later that year.[1] This publication represented our first step in the direction of cosmic biology although we did not recognise it as such at the time.

I soon followed up this first paper with a series of others describing more detailed calculations of the properties of carbon particles, including the possibility of ice mantles condensing around them in interstellar clouds. I also made a prediction of the properties of interstellar grains based on new measurements of the optical constants of graphite, which were brought to my attention by the Belgian astronomer C. Guillaume. The prediction was that if the extinction curve of starlight was extended into the ultraviolet, a strong absorption feature centred at a particular wavelength 2200Å would be seen.[2] It was precisely this feature that later turned to be a spectroscopic beacon for identifying biology even in the most distant cosmic locations.

---

[1] F. Hoyle and N.C. Wickramasinghe, 1962. On graphite particles as interstellar grains, *Mon. Not. Roy. Astr. Soc.*, **124**, 417–433.

[2] N.C. Wickramasinghe and C. Guillaume, 1965. Interstellar extinction by graphite grains, *Nature*, **207**, 366–368.

As a diversion from astronomy in May 1962 I assembled a collection of poems that I had written in previous months and submitted it to compete for an endowed Poetry prize at Trinity College — The Powell Prize for English Verse. Some months later I receive a letter in July from the Senior Tutor to say that I was awarded the prize! This was of course welcome news, but I can only surmise that the competition in that particular year was not very strong.

Throughout most of 1962–1963 my interactions with Fred Hoyle were confined mostly to reporting progress on new developments of the carbon grain story. This was my opportunity to establish myself as an independent researcher which I did with much enthusiasm. Early in 1963 I began writing my PhD thesis and thinking seriously about what I might do when my Commonwealth Scholarship came to an end in September 1963. Our work on the carbon dust grains and on grain formation was opening up new vistas of research in this field, so I did not relish the prospect of returning to a University position in Ceylon, where I may not have been able to pursue my research.

In the autumn of 1962 I decided to apply for College Fellowships in Cambridge. The competition for such fellowships is notoriously stiff. Competitors include young researchers from all academic disciplines, so the best in one field had to be compared with the best in others, a very difficult task for electors, as I was to later discover. So I was delighted when Jesus College elected me to their Research Fellowship in October 1963. This opened a new chapter in my life as a research scientist and a Cambridge don. The Fellowship included a modest stipend, free rooms and meals in College. I moved into my rooms in a new building in North Court to finish my PhD thesis, whilst also taking on a modest amount of undergraduate teaching for the College.

Compared to my somewhat isolated life as a graduate student at Trinity, I began to enjoy the fellowship of my colleagues at Jesus who came from a wide range of academic disciplines. The President of Jesus College at the time was a mathematician Alan Pars, who had taught Fred Hoyle as well as my father in the 1930's. Alan Pars very generously took me under his wing and made me feel comfortable in my new surroundings. Jesus College had no great traditions in astronomy at the time, but had in its history an eminent alumni: John Flamsteed (1646–1719), the first Astronomer Royal of England. Flamsteed's monumental publication of the first extensive star catalogue set the highest standards for observational astronomers who were to follow him. His work was also a great boon to navigation, providing a basis for the accurate determination of position at sea.

In the congenial setting of my rooms at Jesus College I was able to pursue my researches on various aspects of interstellar grains. It seemed clear to me that a major paradigm change was round the corner — a shift from volatile ice grains to refractory grains, grains that had to be based largely on the element carbon. It was also a shift from ideas of grain formation in diffuse interstellar clouds to condensation of dust in much denser stellar environments. The transition, however, was not as smooth and painless as I had anticipated it to be. The publication of our papers on interstellar graphite led to a head-on clash with proponents of the ice grain theory in the USA, particularly Mayo J Greenberg and his collaborators.

The next steps in the progress towards unravelling the precise nature of cosmic dust required observations of stars that extended outside the visible range of wavelengths. New techniques in observational astronomy were now making it possible to study the behaviour of interstellar dust at longer wavelengths beyond red (the infrared) and shorter wavelengths beyond blue (the ultraviolet). By 1965 the absence of an infrared absorption band in the spectra of stars near the wavelength 3.1 micrometre, the diagnostic of water ice, led to the conclusion that ice particles if they exist at all can make at most only a very minor contribution to the interstellar dust.

The observations of stars using telescopes on the ground were affected by absorption of light as it traverses the Earth's atmosphere. Essentially all the ultraviolet light of stars was filtered out in ground-based observations. With the dawn of the space age in the mid 1960's, astronomical observations were possible from above the atmosphere using equipment carried on rockets and satellites. Such observations of stars in the ultraviolet revealed a conspicuous feature in the form of a broad peak of absorption centred on a precisely defined wavelength 2175Å. This was exactly the property we calculated for spherical graphite particles of a particular size — of radius 0.02 micrometres. The old ice grain model of cosmic dust could not reproduce such an ultraviolet feature and must therefore be deemed to be inconsistent with observations and abandoned. This is the way of science, or at any rate it should be so.

The new infrared and ultraviolet observations provided enough reason for a specialist scientific conference on interstellar grains to be convened. Such a conference was promptly organised by J. Mayo Greenberg under the auspices of NASA and held at the Rensselaer Polytechnic Institute in Troy New York from 24–26 August 1965. This was my first visit to the United States and memories connected with it still linger. Landing at La Guardia Airport in New York I spent a day walking around in the shadow of skyscrapers in Times Square. But after the initial bedazzlement the stark reality of American society began to impinge. Concepts of aesthetics and grandeur in the US differ markedly from those of the old world. Here is a society exceedingly self-satisfied, espousing materialism without any self-criticism, dangerously nationalistic. A new President had plunged the nation into full-scale war in North Vietnam in an abstract cause of staving off communism. Freedoms of all kinds including the freedom to protest are enshrined in the constitution, but in many instances its expression is muted. Civil rights for blacks in the southern states had yet to be won.

The next day I arrived at Troy for my first proper scientific conference and I felt a heavy burden of responsibility to defend the graphite grain theory. Mayo Greenberg, the local host of the conference, and the person van de Hulst had referred to during my visit to Leiden, turned out to be my sworn enemy. He was determined to defend the ailing ice grain theory using every trick that was available to him, but it was to no avail. The universe itself is the formidable adversary to defenders of incorrect theories and so it was in this instance as well.

There were a few papers presented at this conference which had an impact on the course of my life. A presentation by Bertram Donn and Ted Stecher from NASA's Goddard Space Flight Centre also dealt with aspects of the carbonaceous model of dust, which was not unlike ours in many ways. Perhaps the most remarkable paper of all was one by a Californian Chemist Fred M. Johnson in which it was argued that a host of "unidentified diffuse absorption bands" in visual stellar spectra can be caused by a derivative of chlorophyll (a porphyrin, the green colouring substance in plants). This paper was several decades ahead of its time in heralding the cosmic theory of life.

A time comes in a journey to step back and take stock of what has thus far been achieved. When I first started my studies in 1960 cosmic dust was considered to be a barren field for research. To most astronomers the presence of dust in the cosmos was a nuisance, its only effect being to hinder the observations of distant stars. All that was needed were simple rules to correct measured intensities of starlight to compensate for the presence of dust, beyond that interest in dust was minimal. In the few years of my research this situation seemed to be changing. Interstellar grains were certainly coming into vogue with new observational opportunities and techniques paving the way to new ideas, and to ambitious programmes of work.

The group of international astronomers who attended the Troy conference in August 1965 were all aware of my work on graphite grains and new avenues and opportunities were beginning to open. Many were sympathetic to the idea of carbon grains, evincing more than a passing interest join me to challenge the ailing ice grain theory. Bertram Donn of NASA's Goddard Space Flight Center was particularly keen in pursuing certain aspects of the carbon dust theory that so far remained unexplained. Some years earlier, he and John R. Platt had argued a case for large unsaturated organic molecules forming in interstellar space. He now felt that a similar process could more easily operate in the atmospheres of cool carbon stars. The work of Platt and Donn in 1960 gave what was perhaps the first hint of large organic molecules occurring in space. It was therefore not surprising that Donn thought it worthwhile to explore further aspects of our ideas on the theory of carbon grains. He accordingly arranged with the Department of Physics and Astronomy at the University of Maryland for me to be appointed a Visiting Professor in the Summer and Fall of 1966. In this way I would be able to interact with his Astrochemistry research group at the NASA Goddard Space Flight Center in Greenbelt which was a short distance from College Park Maryland along the freeway.

But long before my meeting with Donn, Fred Hoyle had arranged, through his friend Willy Fowler, that I spent the fall semester of 1965 at the Kellogg Radiation Laboratory at Caltech in Pasadena. To accommodate both these arrangements I obtained leave of absence from Jesus College Cambridge for the entire academic year 1965/66 to work part of the time in the United States and the rest in Sri Lanka. The Sri Lankan stint was arranged in order to take up an appointment as a Visiting Professor of Mathematics in a newly-formed Vidyodaya University (now Sri Jayawardenepura University). Here I thought I might be able to assess a possible long-term option of returning eventually to Sri Lanka. The research task that I set myself to do during the Sri Lankan spell was to complete a technical book on "Interstellar Grains" that was commissioned by Chapman and Hall (London) in their International Astrophysics Series. This I felt was an important step that would put our new ideas on interstellar grains firmly on the scientific map.

The first phase of my plan for 1965/66 did not turn out to be as successful as I had hoped it would be. Indeed my semester at the Kellogg Radiation Laboratory in Caltech took off to a bad start, when on my first evening, I decided to take a stroll along the wooded streets just outside the campus precincts in Pasadena. Within minutes a police car pulled up beside me and a slow-witted, thick-set cop jumped out to question me as to my business of walking on the street. Everyone, he explained, went by car and walking here was simply against the law. My crime, for which I was let off with a warning, was compounded by the fact that I did not carry any money or form of identity. This incident, which I felt to be an infringement of my freedom, was an unfortunate introduction to life in the United States.

I had expected to be greatly stimulated by the scientific atmosphere of Caltech, one of the most distinguished centres of scientific research in the world. But this was not to be. During my first weeks at Kellogg all my attempts to interest astronomers with problems connected with cosmic dust did not meet with too much success, and this was a disappointment. Most of the astronomers here were preoccupied with "much bigger" problems of cosmology and they had little time for much else. This lack of enthusiasm for problems connected with interstellar grains might sound strange from a modern perspective. Today, in 2014 many space observatories (e.g. the Spitzer Telescope) are dedicated to probing the universe in the infrared radiation emitted by dust grains. Infrared astronomy began to reach maturity towards the beginning of 1968, so my visit in 1965 was perhaps only a few years too early. I left Caltech at the end of November 1965, a little sooner than I had anticipated, and headed home to Colombo, via Hawaii and Tokyo.

Landing at Katunayake Airport with the prospect of a full five-month stretch in my homeland filled me with a sense of anticipation. Compared with the maze of freeways around Los Angeles that I had just left, the journey from the airport to my home in Colombo appeared almost medieval. Pedestrians, cyclists, bullock carts, cars and buses jostled for space along a narrow winding road. Disorderly humanity wound its own path along crowded routes and a twenty-five mile journey took well over an hour and a half.

In a span of five years the city of Colombo as well as the immediate environs of our family home in Bambalapitiya had changed dramatically. The capital had become much more congested and polluted, and one sad consequence was that the magnificent spectacle of the night sky that I had so often enjoyed from my front garden, had now become a rarity. Street lights just outside our house and a more general haze of light pollution had spoilt my ability to enjoy the beauty of the cosmos.

The experience of being installed as a Professor in a new Ceylonese University turned out to be more daunting than I imagined. I found myself being instantly drawn into a great deal of arduous administration. This was perhaps to be expected in a new university institution, but I did not find this part of my work particularly agreeable or rewarding. A poorly stocked University Library made any chance of continuing front-line research in my highly specialised field of astronomy somewhat remote. This would of course have been different had the internet come 40 years earlier! Fortunately I had brought with me all the material that I needed for writing my book on Interstellar Grains. This is what I was able to do in the time that was left for me between my other commitments. And for the rest I simply had to bide my time until I returned to Cambridge, back to my Jesus College Fellowship and eventually to a staff position at Fred Hoyle's new Institute of Theoretical Astronomy. The Institute itself became a statutory reality during my absence, with its buildings in Madingley Road due to be completed in the following year.

By far the most momentous event in my life, one that was to have a profound influence on my career as a scientist, happened during my sojourn in Sri Lanka. This was my meeting with Nelum Priyadarshini Pereira (known to her friends as Priya and to her family as Nel), an earnest law student at the University of Ceylon, and our subsequent marriage in

April 1966. I am often asked whether our marriage was arranged. My answer always is an emphatic *no*. It should be said, however, that those were times in Ceylon that were very similar what had prevailed in Victorian England. Respectable young girls were always chaperoned, and to meet boys or prospective suitors introductions were required. It turned out by a curious quirk of fate that Priya had a close school friend who lived just opposite our house, and my parents and hers had somehow contrived a meeting at this neighbour's house. There was never a question asked by anyone: "would you like to go out with her, let alone marry her?" and I hope a similar question or compulsion was not forced upon Priya on her side. In the event, however, a meeting took place. We enjoyed each other's company and became naturally drawn to each other. After a few meetings I, for myself, must confess that I was smitten! I fell passionately in love with the beautiful, charming and accomplished girl. Within a few months we were married.

So it was that Priya at the tender age of 20, a beautiful, intelligent, determined and talented young girl enters my life. As she still continues to taunt me, she was wrenched from her family and a highly promising legal career (her father was a leading lawyer) to become my wife. As my work in science turned in directions that became ever more controversial, Priya's steadfast support and encouragement, and most of all her combative temperament was a crucial factor in my survival. It might be said with honesty that Priya helped me endure the slings and arrows of outrageous fortune, in the instinctive belief that the ideas so carefully thought out by Fred Hoyle and me must eventually turn out to be right! Little did Priya realise in 1966 that she was stepping in with me into a lion's den of acrimony and controversy in the years that lay ahead. Such is the nature of science, when it comes to challenging and contesting accepted views and dogmas. But the worst of all that was still a full decade into the future.

In April 1966 I went on the third sea voyage of my life, this time in the company of Priya, from Colombo to Southampton, bidding farewell to family and friends, and the country of our birth. My sail away from Colombo on this occasion was more poignant than on earlier occasions that I have described. This was in a real sense a decisive farewell to our earlier life in the midst of our families and friends. The wrench must have been far more traumatic for Priya as this was her very first separation from her family, particularly from her sister Ramani. Priya was especially close to her sister so the pain of parting was strongly felt. Ramani Pereira remained in Sri Lanka to become our lawyer in the family and she still continues to work as the Chief Legal Consultant for the Times group of newspapers. Now, with modern modes of communication — internet, mobile phone, *Skype* — the two sisters keep in close touch, but when we first left the island in 1966 a physical separation meant an almost total loss of contact, and this would have been difficult to bear.

Returning to my story, Priya and I arrived together in Cambridge to set up our first house in a flat in Jesus Lane that was provided for us by Jesus College. After two months in Cambridge we were on our travels again. This time by ship *SS France* to New York, and then on to College Park, Maryland where we would be based for three months. Although Priya and I agreed that we could never contemplate living full-time in the United States, we made the most of our stay here, spending our weekends touring the exquisitely beautiful and well laid-out national parks of Virginia, Baltimore and Washington in our old Chevrolet that we had bought for $50.

My office on campus was located in the Department of Physics and Astronomy headed at the time by the Dutch radio astronomer Gart Westerhout. My appointment as a Visiting Professor was in the University of Maryland, but since the funding for my post came from NASA, most of my collaborative links were with Bert Donn's Astrochemistry group at the Goddard Space Flight Centre in Greenbelt. It took me a while to get accustomed to the new environment and particularly to the work ethic that prevailed. I had got used to working Cambridge-style on an individualistic basis, even in my collaborations with Hoyle. Here at Goddard I was drawn into more restrictive communal research atmosphere. People liked to talk a great deal and to work, or appear to work in big teams. There was always an impression of great industry, with large numbers of people working at a frenetic pace on a single problem. But the output was not always commensurate with the manpower or effort that was expended. At least that was the way it seemed to me in the mid 1960's.

As I had agreed with NASA scientist Bert Donn my task in the three months at Maryland was to investigate further the problem of the formation of carbon grains. The work as I remember it involved a tiresome succession of meetings and conferences occupying a great deal of time.

And even at the end of my stay a final manuscript for submission was not agreed upon. Many letters and drafts had to be exchanged across the Atlantic, and the ultimate result was the publication two years later of a paper in the prestigious journal Astrophysical Journal.[3]

American scientists in 1966 were clearly ahead of the British in the practice of counting research papers and citations in a race to justify their existence, and to secure continued public funding for their projects. The levels of stress generated by this process were visibly detrimental to the health and well being of the scientists concerned and also to the progress of science. Nowadays we have come to take this practice for granted, with universities and research groups vying to get the maximum number of papers published in the shortest time in the so-called high impact journals, and the maximum number of citations. Our papers on interstellar dust published throughout the period 1962–1968 and beyond were all in high impact journals, and they certainly had a profound effect in helping to change astronomical fashions from volatile ice grains to refractory grain models.

Priya and I developed a passion for travelling and exploring the world very early in our life together. After our brief spell at Maryland in 1966 that I have just described, our next big trip was to Japan in 1969. The assignment was as a Visiting Scholar at the famous Yukawa Institute in Kyoto for three months and we set up house this time in a Kyoto University apartment within walking distance from the Institute. I gave lectures and seminars on interstellar dust and interacted mainly with Haruyuki Okuda at the time a young lecturer at the University. Haruyuki and his wife Ikuko became family friends and we have kept in touch ever since. Another Japanese friend we re-connected with during this visit was Shin Yabushita, who was a PhD student working with Ray Lyttleton in Cambridge in 1961.

Japan in 1969 was an electronic dreamland — the birthplace it seemed of many new technologies. I loved browsing in the many electronics and camera shops in Kyoto as well as Tokyo inspecting their bewildering range of new gadgets. With a common culture based on Buddhism, Priya and I acquired an empathy with the Japanese people and a love of Japan that has endured throughout our lives.

---

[3] B. Donn, N.C. Wickramasinghe, et al, 1968. On the formation of graphite grains in cool stars, *Astrophys. J.*, 153, 451–464.

CHAPTER 9

THE DATA LEADS THE WAY

As I have already mentioned, my assignments as a Professor at Vidyodaya University and later as a researcher at Caltech and Maryland were taken up during a year's leave of absence from my Fellowship at Jesus College. On my return to Cambridge I resumed my Fellowship but also took up an appointment as a Staff Member of the newly formed Institute of Theoretical Astronomy at the University. My fixed-term Research Fellowship was simultaneously converted to a normal Fellowship of the College, and I was also appointed as a Tutor and a College Supervisor in Mathematics. Each Tutor is assigned a small group of undergraduates to whom he is supposed to serve in a type of pastoral role, *in loco parentis*. Entertaining students once a term to sherry parties and being responsible for their general well being whilst at University was a new task for me and an experience I began to enjoy.

The setting-up of an Institute of Theoretical Astronomy in the University (to which I was now appointed) had been Fred Hoyle's ambition for several years. Its formation freed him (and his colleagues) from some of the constraints had we continued our traditional attachment to the Faculty of Mathematics. It had been a hard struggle for Fred Hoyle to convince the University of Cambridge as well as a private funding agency (the Nuffield Foundation) to support such a venture, but eventually the battle was won.

In summer of 1967 the Institute of Theoretical Astronomy in Cambridge came into physical existence in a well-equipped open-plan building in the midst of a meadow off Madingley Road. It seemed to be strategically located between two friendly institutions — the Cambridge University Observatories on the one side and the Geophysics Department on the other.

Fred Hoyle's cosmological adversaries at the Mullard Radio Astronomy Observatory were located only a couple of miles away on Madingley Road, but as far as interaction was concerned they may as well have been as far away as the Moon. Martin Ryle and his team were continuing in their single-minded pursuit of disproving Steady State Cosmology. From their studies of the counts radio sources to various intensity levels, they claimed that radio-emitting galaxies appeared to be closer together as one goes to greater and greater distances in the Universe, showing that the Universe could not be in a steady state.

The real crisis for Steady-State Cosmology was, however, not radio source counts but the new discovery by Arnold Penzias and Robert Wilson of a cosmic microwave background with a temperature of 2.73 degrees above absolute zero, identified as a relic of the Big Bang. To offer any credible defence of Steady-State Cosmology it was imperative to explain this background by a process that was unconnected with an early hot phase of a Big Bang Universe.

The conduct of creative pursuits through the course of one's life cannot be disconnected from what is happening in the wider world. Just as I was beginning to feel settled with Priya in our new life in the UK racial tensions began to grow on both sides of the Atlantic. On April 11th 1968 Civil Rights leader Martin Luther King was gunned down in Memphis and a wave of race riots spread across major US cities. Within an amazingly short space of time ripples of racial disquiet reached Britain. On April 21st Enoch Powell made his historic "rivers of blood" speech. Powell, a distinguished classical scholar, made his point most eloquently thus: "As I look ahead I am filled with foreboding. Like the Romans I see the River Tiber foaming with much blood." He went on to say that Britain must be, "mad, literally mad as a nation to admit 50,000 dependents of immigrants into the country every year". The present situation he concluded is like a nation "busily engaged in heaping up its own funeral pyre." Conservative Party leader Ted Heath was quick to denounce Powell's speech and expel him from the shadow cabinet, but for those of us who were attempting to adopt Britain as our home, these developments were a source of extreme anxiety and insecurity. I shall touch on matters of racial prejudice as it related to my own life in the United Kingdom in later chapters indicating perhaps that Enoch Powell was right after all. Being thoroughly engrossed in my research work was a solace and a distraction from the growing disquiet in the world around me.

The summer of 1969 had a special significance for Priya and myself. Fred Hoyle's wife Barbara visited Ceylon with her friend Viv Howes, and we had the pleasure of looking after them. Whilst they were in the capital they stayed in my parents' house but they spent many days touring the island. We accompanied them on most of their trips and together with our visitors we saw things and places we had never seen before. The 21 years of our lives that we spent in Sri Lanka were evidently not enough to have explored all the remarkable sites of this island. Our trips took us to the ancient ruins of Anuradhapura and Polonnaruwa, the hill country capital Kandy, to tea plantations and to the beaches on the West coast of the island. Whilst in the hill-country station of Ella, the place that made the greatest impression on all of us was "Land's End". In the first morning mists at the guest-house where we were staying, we experienced that peculiar sense of the infinite, of our dwelling here on Earth giving way to the sky and the eternal infinite cosmos. Gigantic stone statues of the Buddha set among the arid planes of central Sri Lanka reinforced this feeling. Anita (infinity) lies at the heart of the Buddhist philosophy — a world without end or limit, into which the individual may at last be merged.

Back in the world of science dramatic events were unfolding following Neil Armstrong's iconic walk on the Moon. There dawned a new sense of determination to make progress in space research and space exploration, with NASA having regained its supremacy over its Soviet rival. A new generation of telescopes and instruments carried aboard satellites came into full operation. For example NASA's Orbiting Astronomical Observatory 2 (OAO2) began to yield a wealth of new data on ultraviolet spectra of stars that had a direct bearing on the composition of interstellar dust. The existence of a conspicuous ultraviolet absorption feature of interstellar dust centred at about 2175Å (to which I have referred) was confirmed. A striking result that emerged from the new studies was the invariance of the 2175Å absorption hump from star to star, thus declaring the universality of some chemical component of the cosmic dust.

This invariance of the ultraviolet feature, according to our original carbon grain models, demanded the presence of graphite particles in the form of spheres with radii fixed almost at a definite value, 0.02 micrometres.

Even as early as 1969 we were beginning to feel uneasy about the artificiality of these assumptions. Graphite (which is mined in large amounts in Sri Lanka) occurs in the form of microscopic flakes, not spheres. The "lead" in lead pencils is not the metal lead, of course, but cylindrical tubes of compressed graphite. When you write with a lead pencil these microscopic flakes of graphitic carbon are transferred to the sheet of paper. So a little thought tells us that the assumption of minute spheres of graphite all of the same size is artificial and cannot be correct. But a properly formulated realistic alternative to graphite spheres in space, and indeed a connection with life, lay a few years into the future.

After our return to Cambridge we began to feel more settled in our lives at least for a while. In November 1970 we had our first child Anil, who 17 years later was to read Mathematics at Jesus College Cambridge. The period 1967–1970 saw also the emergence of the new discipline of infrared astronomy. Exploration of the universe at wavelengths that were longer than visual wavelengths began in earnest. The developments of new infrared detection techniques and the use of ground-based telescopes at high, dry mountain sites, provided a wealth of new information about the cosmos. A new window in the electromagnetic spectrum was open to observe the universe.

The first discovery of infrared astronomy directly relevant to my story came from the work of John E. Gaustad and his colleagues who confirmed that the infrared spectra of highly reddened (dimmed) stars showed no evidence whatsoever of water ice. This was a victory for the carbon dust theory and a setback for the old ice grain theory. Next there followed a spate of discoveries all showing the presence of a new spectral feature of dust over the infrared waveband 8–12 micrometres. The feature was observed in emission in a wide variety of astronomical objects and it was immediately interpreted as evidence that the cosmic dust was made of a mixture of silicates — combinations of magnesium, silicon and oxygen as they occur in the rocks of the Earth.

When I came to critically review this interpretation, what struck me immediately was how poorly the newly observed astronomical feature at 10 micrometres actually fitted the behaviour of any known silicate or mixtures of silicates. Because mineral silicates (e.g. rocks) do indeed have absorptions spanning the 8–13 micrometre waveband, there was a crude match to be seen. But the identification of silicates in vast quantity in interstellar space was by no means compelling at this point. Other chemical systems, including some that involved carbon, could be stronger candidates. Whilst one could not dispute that some quantity of silicate dust might exist in space, how would this compare with contributions from the far more cosmically abundant element carbon? This was the question that I explored for over a decade before arriving at the thesis that complex organics and even biological material might be involved in producing the infrared signatures that were wrongly attributed to mineral silicates. My second clash with astronomical orthodoxy was now to begin.

Our search for a possible carbon-based interpretation of the 8–13 micrometre interstellar feature began as early as 1969. An examination of spectroscopic literature showed that soot containing admixtures of hydrogen had spectra that might be satisfactory from this point of view. We were, however, aware of some glaring shortcomings. The fit of our synthetic model spectra to the new mid-infrared data, although better than for silicates, still left much to be desired. For the quality of the observational data that was now available a more perfect agreement with models was required. There was also the related question I referred to earlier, of justifying the artificial requirement of perfectly spherical graphite particles all of one size.

With a view to resolving at least some of these questions I began to explore models of cosmic dust comprised of three separate components — mixtures of graphite, silicate and iron particles. Although fits to the astronomical data were found for such a multi-component model, relative proportions had to be fixed arbitrarily, and this was a source of concern. We were approaching a situation analogous to that which prevailed in the middle ages when there was a struggle to defend Earth-centred Ptolemaic models of the solar system. Every new observation called for a new "epicycle" and it was a similar situation we now had in relation to models of cosmic dust. It is curious that precisely the models that I reluctantly proposed for the first time in the early 1970s are now accepted without dissent. We ourselves eventually moved away in the direction of biological models as we shall see. But before we plunged into the depths of outrageous heresy there were the fruits of victory to be reaped.

Ten years after the publication of our seminal paper on graphite grains, and with the mounting evidence against the old ice grain theory, I was much sought after as a plenary lecturer at conferences on interstellar material. I was flattered to think that I had replaced the great H. C. van de Hulst as one of the reigning experts in the field of cosmic dust. In March 1972 I had the good fortune to be invited to be the principal lecturer in an Advanced Winter School on Astronomy and Astrophysics held in the famous Swiss Ski resort of Saas Fee. The theme of the School was "Interstellar Matter" and the other lecturers were F. D. Kahn and P. G. Mezger. My own lectures were on Interstellar Dust, in which I described my new theories to graduate students.

The venue of the meeting at Saas Fee was situated in a high plateau in the Saas Valley in the Swiss Alps at an elevation of nearly 6000 feet. The lectures took place only in the mornings so the afternoons were free for walking on the snow clad slopes and skiing if one so wished. I was accompanied by Priya on this trip, as always, but with our young son Anil who was under two years old we were somewhat limited in what we could do. However it was a most memorable respite for us before the next storm was to blow.

CHAPTER 10

FROM CAMBRIDGE TO CARDIFF

The date 22nd May 1972 will go down as an important date in the annals of the history of Ceylon. My native country severs its last formal ties with Great Britain and declares itself a republic, with Mrs Sirimavo Bandaranaike as Prime Minister. From henceforth, reference in this book to this country ceases to be Ceylon, but Sri Lanka — which in Sanskrit means "the resplendent isle".

Momentous changes are also taking place in Cambridge. Within less than a decade of its inception the Institute of Theoretical Astronomy in Cambridge had acquired the status of a Mecca in the subject. There was a steady stream of astronomers from all over the world coming here to meet, brainstorm and collaborate, rather like the great centres of astronomical science that would have been around during the Renaissance in Europe.

Despite the enormous success of Fred Hoyle's Institute of Theoretical Astronomy, in a brief span of four years we were told that its very continuity was being threatened in the summer of 1971. Fred Hoyle relates this part of the story in great detail in his own autobiography,[4] so I shall not reiterate the details, but merely say that things turned out to be so intolerable for him that he submitted his resignation to the University in January 1972.

The news came to us as a shock when it finally reached us in the early Spring of 1972. Cambridge at this time was looking at its seasonal best. The colleges were shedding the austerity of their winter identity and the backs were now softening to a new ministry of snow-drops, crocuses and

---

[4]Fred Hoyle, 1994. *Home is Where the Wind Blows*, University Science Books, Mill Valley California.

daffodils. Stone courtyards, grey too long, glimmered again in sunlight, revealing their glorious red and purple hues. It seemed a cruel fact that ill winds could blow through so exquisite a setting.

The ten years I spent as a Fellow of Jesus College were perhaps the most rewarding of my life in many ways. Not only did my research consolidating theories of interstellar grains give me a sense of personal achievement, but the Collegiate life was also rewarding. I formed lasting friendships with many interesting people amongst the Fellowship. Apart from Alan Pars whom I have already mentioned, others I remember with affection include the Welsh novelist Raymond Williams, and biologist and ornithologist W.H. (Bill) Thorpe. I remember the many discussions I had with Bill from which I learned a lot about biology and also his views on how a belief in God and Science could be reconciled. In my first year as an unmarried Fellow living in College, I dined in hall on a regular basis, and was thus able to get to know many of the Fellows.

After I returned to Cambridge with Priya as a married Fellow I did not dine in hall more than once a week, but the lavish feasts, the Ladies' Nights, *etc.* were certainly events that I would not miss. In those days there were no women fellows or women students at Jesus. Only for one feast in the year, the Ladies' Night, could a Fellow of the College introduce a lady guest. Now of course it is all very different. Equality prevails on all fronts, or so it would seem.

I spent most of the summer of 1972 in Sri Lanka with Priya and our family going almost every weekend to some delightful beach resort on the west coast of the island. The era of mass communication had not yet dawned — faxes, emails and the internet were nearly two decades away and international dialling was expensive. So this tiny island in the Indian Ocean seemed blissfully but strangely remote.

From an early age the sea held an irresistible attraction for me. This is true also for very many people. The theory that life sprang from oceans, accounting for humanity's fascination with the sea, I was later to refute. Whilst Fred Hoyle would always favour the exigencies of rugged landscapes as a means to reflect and meditate, I would invariably choose the ocean where communion is instantaneous. Wave after wave rolled forward from the distant ocean, interrupted only by the occasional rumbling of a train passing behind me. And vexations would momentarily recede, drowned by the great swell of the sea. Every evening the sun would set around 6.30pm. The splendour of sunset that follows is brief because we are so close to the equator. The sunset colours depend very much on the disposition of the clouds and prevailing conditions in the atmosphere. Sometimes the sunsets are spectacular, at other times a disappointment, mirroring the vicissitudes of life.

My thoughts this summer were mingled with nostalgia for my days at Cambridge that I feared were drawing to a close. It seemed also that my collaboration with Fred Hoyle may have come to its logical end and I kept wondering what shape my career might thereafter take. Priya was pregnant with our second child, so thoughts about the future were even more resonant.

When I returned to Cambridge that October the reality of the situation dawned on me even more powerfully. As I walked into the Institute I felt a sense of sadness sweep over me, like returning to a house of bereavement — its life and soul had departed leaving what seemed an empty shell. Hoyle had left the Institute in mid-August, never to return, and shortly afterwards they sold their house in Clarkson Road. Fred and Barbara Hoyle had entered the next phase of their lives and were now in the process of resettling at Cockley Moor, near Pernrith in Cumbria in their beloved Lake District.

In our family life October 1972 saw the birth of our elder daughter, Kamala, who grew up to acquire a great flare for English literature that took her many years later to Jesus College to read English. She has recently collaborated with me on several writing projects, acting in the important role of critic and editor.

My own appointment at the Institute was assured only until the end of the academic year. What was to happen after that would be decided upon by the next Director. Without Hoyle the Institute had lost its attraction for me and I could not see myself carrying on there, even if I were given the chance. The option of looking for openings in a Sri Lankan University was one I had considered but did not seriously entertain for too long. I was still fired with enthusiasm to carry forward the many promising lines of research I had begun, and the scientific culture of Sri Lanka at the time could not permit the funding of such ventures. The country had more pressing economic and social problems to grapple with.

I then began to look for university jobs elsewhere in the UK. One of the first that came to my notice was the Chair of Applied Mathematics and Mathematical Physics at University College, Cardiff. Fred had not yet left the UK astronomical scene, so I was able to consult him on this matter. He strongly advised me to apply. I did so and my application was successful. At the interview for selection the Principal of University College Cardiff, Bill Bevan, made it clear to me that he was very keen to start Astronomy in Cardiff, and furthermore that he would like me to involve Fred Hoyle in this venture. I could not have expected a more attractive offer, so I accepted the position without further ado and agreed to assume duties in the summer of 1973.

Early in the summer of 1973 we sold our house in Barton Cambridgeshire and moved to Cardiff with our two young children. Our new home in Lisvane, a suburb of Cardiff, overlooks gently undulating mountains — a change from the oceans of Sri Lanka I grew up with, and from the flatness of the Fens around Cambridge. At the new University I felt at first a little overwhelmed by the responsibilities entrusted to me. The task was by no means easy. There was jealousy and hostility of the existing staff of the department to contend with, particularly so as I had set out to steer the department in a direction that appeared quite alien to them. However, with the unstinting support of Principal Bill Bevan and senior members of the Senate I was able to institute great innovations.

Within a year of my appointment I had changed the name of the Department from "The Department of Applied Mathematics and Mathematical Physics" to "The Department of Applied Mathematics and Astronomy" and we were soon on the way to appointing four new lecturers in Astronomy. I had a dilemma of deciding upon the research fields in which to choose the new lecturers. Had I opted to have all appointments in the area of interstellar dust we would probably have had the most powerful research group in the world in my own specialist field. Following the example of my mentor Fred Hoyle and the experience of the Institute in Cambridge I opted instead to develop as diverse a group as I possibly could.

The first appointee was in the area of plasma physics, the second in the chemical evolution of galaxies, the third in star formation theories and the fourth in relativistic astrophysics. A year later I obtained support for a Chair of Observational Astronomy, so a highly diverse and balanced group was begun. The choice as I saw it in 1973 was like the difference between sowing a packet of mixed flower seeds and one of a single species of flowering plant. The mixed seeds if they took root would lead to a glorious splash of colour. This was what happened in Cardiff, leading to the evolution of an astronomical research centre that is one of the best in the UK. Such heterogeneity would be unheard of in the present day as subjects have become so overly specialised that people from different fields barely speak the same language as each other, and often cannot recognise that they are trying to address the same problems.

With such far reaching changes I have put in effect and with the unfailing support of Principal Bevan, there were seeds of antagonism that also began to sprout with a vengeance. I had to face the hostility of the old guard in the Department who felt that they were somehow betrayed. Racially motivated animosity also reared its ugly head and overt discrimination with explicit threats against me continued right up to my retirement in 2006 from my full-time Professorship and even beyond. I shall discuss these more explicitly later, but what irks me most was that successive administrations of the university have deliberately chosen to discriminate against me, totally ignoring my efforts in starting the astronomy research in Cardiff that has since become a jewel in its crown.

By the late Spring of 1974 I had set all the major changes of the Cardiff astronomy project in train and was headed again to North America for a short respite — this time to Canada and to the University of Western Ontario in London, Ontario. Priya and I instantly took to Canada finding it much more agreeable than its neighbour USA. The country was more akin to Europe in terms of its culture and value systems, thus we found it more recognisable and comfortable to live in. It was during my 3-month stint in Canada that I made a breakthrough that was to become a defining moment in my research career. I mentioned earlier that I was not entirely satisfied with the quality of fits that had been obtained to astronomical observations over the 8–13 micrometre infrared waveband with silicate dust grain models. So I was still searching for a better solution.

What if the carbonaceous component of the dust was not simply graphite as we proposed in 1962 but made of organic materials, organic polymers in fact? Perhaps carbon atoms in interstellar space might be combined with hydrogen and oxygen to form an extraordinarily vast variety of organic chemicals. In terms of the basic chemical elements at least there would be more than enough mass to explain the properties of interstellar dust.

At this time the molecule formaldehyde H_2CO had been discovered to exist ubiquitously in interstellar clouds. It was present in dense molecular clouds as well as in the less dense interstellar medium. What if such molecules started to condense and polymerise on the surfaces of pre-existing graphite or silicate grains expelled from stars? Looking through the books in the library of the University of Western Ontario dealing with properties of formaldehyde, I soon discovered that it could readily polymerise on the surfaces of silicates. It also turned out by a stroke of luck that the Chemistry Department here had one of the leading experts on formaldehyde polymers and I had the benefit of many discussions with him. A simple calculation showed that under interstellar cloud conditions substantial mantles of formaldehyde polymers would indeed grow. Next I began to look at the optical and infrared properties of many types of formaldehyde polymers. It turned out that such polymers were dielectric (that is to say, non-absorbing) in the visual waveband as the astronomical observations demanded. And most strikingly, polyformaldehyde (polyoxymethylene) had absorption bands over the 8–12 and 16–22 micrometer wavebands, with the former absorption fitting the astronomical data better than any known silicate. This was a breakthrough moment and within a couple of weeks my paper entitled "Formaldehyde Polymers in Interstellar Grains" was submitted to *Nature*. When it was later published in *Nature*[5] it predictably made a major splash in the science news columns of the broadsheets. This was the first-ever suggestion of the widespread occurrence of organic polymers in the galaxy, and it was also the first paper to come from the newly reconstituted Department of Applied Mathematics and Astronomy in Cardiff. It marked the beginning of the cosmic life theory that was developed throughout the 1970's and 1980's.

Within days of the publication of this paper Professor V. Vanysek of Charles University Prague came to visit me with a proposition to be considered. Polymerised formaldehyde and also other organic polymers, could

---

[5] N.C. Wickramasinghe, 1974. Formaldehyde polymers in interstellar space, *Nature* 252, 462–463.

form a major component of comet dust as well. Comets were thought of at this time to be dirty snowballs, mostly comprised of inorganic ices with siliceous and metallic dust occurring as minor impurities. When a comet approaches the inner parts of the solar system, the molecular species in the developing coma were thought to be broken-up products of ices which were regarded as "parent molecules". With polymerised formaldehyde being a component of the comets, coma radicals such as OH, CN, C_2 could be interpreted as the broken-up products of such a polymer. When we began to explore this idea together it appeared to us more and more plausible. Since formaldehyde polymers remain stable up to high temperatures of 500 Kelvin (227°C), changes in the 10 micrometer spectrum of the new Comet Kohoutek (1973f) as it came within 0.5AU of the sun (1AU = the average distance between the Earth and the Sun) were shown to be consistent with this organic model. The idea of an organic comet was thus born, and its justification was described in a joint paper.[6] This publication establishes our priority for the idea that both interstellar and cometary dust could be comprised predominantly of organic polymers.

Up to this point in my story there was no major discord with mainstream astronomical ideas, and indeed there were even some grudging plaudits being accorded to us for making such an innovative suggestion. I soon acquired a research student, Alan Cooke, who started doing more detailed modelling of formaldehyde grains, and a grant from SERC (Science and Engineering Research Council of the UK) to do experimental work on the optical properties of organic polymers was also secured. From 1975–1977 I was awarded more grants from SRC that enabled several visitors to be invited to Cardiff for collaborative work. They included Donald Clayton, Donald Huffman, Craig Bohren, Asoka Mendis (my old University friend from the USA) and D. P. Gilra, Jayant Narlikar, S. P. Tarafdar and Pushpa Joshi (from India) amongst others. In a small way the Astronomy activity in Cardiff was beginning to look like a miniature version of Fred Hoyle's Institute of Astronomy in Cambridge. But just as with all the good things of life that position was destined not to last.

---

[6]V. Vanysek and N.C. Wickramasinghe, 1975. Formaldehyde polymers in comets, *Astrophys. Sp. Sci.*, 33, L19–28.

CHAPTER 11

FROM ORGANIC DUST TO ECODISASTERS

In the year 1974 the 40-metre Anglo-Australian Telescope, a unique instrument of its kind at the time for observing the southern skies, was formally inaugurated by HRH Prince Charles. Fred Hoyle had played a major role in the realisation and completion of this project on the UK side. When UK funding for the project was in the balance it was Fred's powers of persuasion and personal intervention with the then Minister of Science, Margaret Thatcher, that eventually clinched the deal. A decade later, when Margaret Thatcher was Prime Minister, the Prime Minister of Sri Lanka, Premadasa, was on a State Visit to the UK. Priya and I were fortunate to be invited as guests at a luncheon at No. 10 Downing Street hosted by Margaret Thatcher in honour of the visiting Prime Minister. I spoke at length to Mrs Thatcher on this occasion and our conversation naturally turned to Fred Hoyle and the AAT. Mrs Thatcher recounted a story later confirmed by Fred Hoyle. When it was her duty to defend the UK's share of the expenditure for the Telescope to the government she had to present a case, and she consulted Fred Hoyle for advice. Fred had told Mrs Thatcher to simply state this: "When a major TV programme on Astronomy is aired on the BBC it has the highest viewing figures." To which Margaret Thatcher simply replied: "Say no more", and of course the AAT was funded and came into existence. As we shall see later infrared observations carried out using the AAT by my brother Dayal and David Allen provided crucial evidence that supports the cosmic theory of life.

Two years after I began my tenure at University College Cardiff, Fred Hoyle was appointed as an Honorary Professor. This meant that he would be a nominal member of my Department and be entitled to use the university as endorsement for his publications. In turn, Cardiff gained much kudos from this link with Sir Fred Hoyle, who was an iconic symbol of 20th century theoretical astronomy.

During the period of 1975–77, Fred Hoyle spent large chunks of his time in the United States, so that his appearances in Cardiff were rare. Even so, I usually managed to track him down and keep him abreast of our research developments on organic polymers, which was proceeding with unprecedented vigour at the time.

Most of my research efforts were now directed to finding the best possible fits to the infrared emission feature of cosmic dust over the infrared waveband 8–13 micrometre. I concentrated in particular on one astronomical source — the Trapezium region of the Orion Nebula where heated dust emitted infrared radiation over precisely these wavelengths.

There was now what seemed to me to be unequivocal evidence of organic polymers existing on a vast scale throughout the galaxy. Models involving co-polymers — mixed chains of formaldehyde and other molecules as well as polymer mixtures resembling tars — were leading inexorably in one direction — life. What if the interstellar grains that I had begun investigating in 1960 were indeed connected with biology, with life itself? This question, with all its profound implications, represented such an assault on conventional thinking that I felt a compulsion to involve Fred Hoyle at this stage.

I began a correspondence with Fred Hoyle early in 1976 by first suggesting that the polymeric grains in molecular clouds could represent the beginnings of a process that may lead to life, thus permitting life to originate and evolve on a much bigger scale than had hitherto been contemplated. Fred's first reactions were far more cautious than I had anticipated. This point is worth stressing because there is a general perception that he embarked on our joint projects rashly and uncritically. Nothing can be further from the truth. He was exceedingly critical of every radical proposition that was put to him at each stage in our collaboration. He played the role of devil's advocate until he was convinced that there were overwhelming arguments to support the radical proposition. And this is exactly how a true scientist should proceed.

After many exchanges of letters and papers Fred Hoyle and I agreed in August 1976 that organic polymers in grains could undergo a Darwinian-style prebiological or prebiotic evolution during the collapse of a molecular cloud in the process of forming new stars and planetary systems. Organic tarry grains tend to be sticky and grain clumps would form by particles colliding and sticking together. Such grain clumps would also trap other organic molecules from the gas, and chemical transformations, sometimes assisted by ultraviolet light, would take place in the condensed state. In our first paper published in the journal *Nature*[7] referring obliquely to biology we wrote:

"The formation of simple amino acids (e.g. glycine) is expected to take place in dense molecular clouds which may well be the cradle of life."

Even such a tentative proposition was regarded as outrageous heresy in 1976, although nowadays it is regarded as obvious.

Towards the later part of 1976 there were other events that were to have a bearing on our story. Even at the risk of digressing from the main thread of my argument I shall report one such event here if only to keep my record in an approximate chronological order.

Space exploration was gathering momentum, including the search for life elsewhere in the solar system. There were two unmanned missions to Mars: Viking 1 arrived at the red planet on 20 July 1976 and Viking 2 on 3 September 1976. Each mission involved an "orbiter" that was set in motion around the planet and a "lander" that actually set down at a chosen spot. The two Viking landers arrived safely at their chosen destinations, equipped with apparatus to make *in situ* tests for the presence of microbial life. These experiments were of vital importance and in many ways much more explicit in detection of life than any subsequently conducted experiments.

In one experiment (designated LR) a nutrient broth (with an isotope label on the carbon) of the sort that is normally used to culture a wide range of terrestrial microbes was contained in a sterilised flask, and the Martian soil was robotically added to it. It was found that the nutrient was taken up by the soil and gases (CO_2) were expelled from the flask as would be expected if bacteria were present. In another experiment the soil sample was heated to 75 °C for three hours before it was added to the

---

[7]F. Hoyle and N.C. Wickramasinghe, 1976. Primitive grain clumps in carbonaceous chondrites, *Nature*, 264, 45–46.

nutrient. This led to a diminution of gas release by 90%, but significantly the reaction was not completely stopped. Since some bacterial and fungal spores could survive temperatures of 75°C, the result of the second experiment was also consistent with a biological explanation, especially as the activity recovered gradually to its former higher value as time went on. The bacterial explanation gained further support from a third result, obtained by heating the soil sufficiently to kill microbial life entirely, when all activity was found to stop. However, another experiment in the Viking package proved initially more difficult to reconcile with biology. This experiment designated GCMS, sought to analyse the organic content of Martian soil using a mass spectrometer. Here the results were disappointingly negative for organic matter, indicating that if such matter existed it was present only in the most minute of quantities.

The fact that the LR experiment was decisively positive and the GCMS experiment was negative posed a difficulty for NASA. The outcome was indefinite, and this is the way it should have been presented to the public. Yet NASA elected in 1976/77 to announce that the Viking experiments did not support the presence of life, and their statement that Mars was an intractably lifeless planet was given a great deal of publicity in 1976. It was their view that some other non-biological explanation had to be sought and would eventually be found. This has not happened to date and in a re-examination of all the Viking results in 2012 it has to be conceded that the balance of evidence is strongly positive for Viking landers to have detected life. But even retrospectively, there still remains an enormous resistance to admit this, and modern pronouncements by NASA relate mostly to the presence of past life in an epoch when rivers flowed over the surface of the planet. The facts have been clouded over, perhaps in an attempt to hide the fact that they plumped for an incorrect theory in the first instance. And when big government science and the media make errors together, neither is anxious to be seen correcting itself, a sufficient reason why nothing much that is good can come of public funded science done in the glare of publicity. Science in my view is a quiet, reflective activity, which cannot flourish in modern egalitarian or totalitarian societies.

The principal investigator of the 1976 biology experiment on Viking, my friend Gill Levin, has revealed many things that are not generally known to the public. For instance his studies of a sequence of pictures taken by the Viking cameras over the duration of a Martian year, showed subtle shades of green appearing on the tops of rocks in the spring. These receded in the winter, suggesting the growth of lichen-type microbial life. Levin's views about all these findings, however, did not endear him to the administration of NASA, and he parted company from them shortly afterwards to pursue his investigations independently. And after several decades of experimentation it has turned out that no non-biological model is feasible for explaining the positive results of the Viking LR experiment, and moreover the lack of free organic matter in quantity as revealed by the GCMS experiment can easily be explained on the basis of a slow turn-over rate of microorganisms to be expected under the relatively inhospitable conditions that prevail on Mars. It is interesting that when a prototype of the Viking lander experiments was taken to the dry valleys of the Antarctic, very similar results to those obtained on Mars were obtained. We know of course that microorganisms exist in abundance in these regions on Earth, so the conclusions drawn from the 1976 Viking experiments on Mars are patently flawed.

The Mars probe Odyssey was launched in April 2001 to orbit the planet and map its surface for hydrogen, water and minerals. Named *Odyssey* after Arthur C. Clarke's blockbuster novel, the probe obtained pictures that showed clear evidence of heavy frost or snow in many locations including the Viking landing sites. Snow or frost deposits were found to be seasonal, pointing to some kind of water cycle. But still NASA were claiming that contemporary life was highly improbable, despite the fact that on many sites on Earth where life has been discovered — in Antarctic ice and at depths of 8 km below the Earth's surface — there are unquestionable parallels with Mars. In 2004, the spacecraft "Mars Express" obtained traces of methane and oxygen in the atmosphere that together would normally be interpreted as indicating biological activity. NASA seems loath to publicise such findings, perhaps in the hope that if they present the case very slowly, they would then be able to claim exclusive priority for the discovery of extraterrestrial life at some future moment in time.

In 2013 NASA's Mars Curiosity Rover has reported a positive detection of complex organics at the lander site. Taken together with the labelled gas release results from Levin's 1976 experiment on Mars, Gil Levin has justifiably claimed that the presence of microbial life on this planet must be taken as irrefutable fact.

Arthur C. Clarke summarised the four stages of the way new ideas are accepted into mainstream, institutional science:

(1) These ideas are crazy, don't waste my time with them
(2) These ideas are possible, but are of no importance
(3) We said these ideas were true all along
(4) *We* thought of these ideas first

I shall now return to the story of interstellar dust. I had been plagued for a while by the thought that graphite grains, for which idea I had by now acquired a degree of fame, could not offer a rational explanation of the 2175Å interstellar absorption band. The requirement of spherical graphite grains, all of one radius, 0.02 micrometres, was difficult if not impossible to defend. The central wavelength of the absorption due to graphite shifted away from the astronomically observed wavelength for particles in the shape of flake or whiskers, or if the radii of spheres departed significantly from a finely-tuned value of 0.02 micrometres. Whilst continuing to refine correspondences with the infrared spectrum of the "Touchstone source" the Trapezium nebula, with various types of organic polymers, it occurred to me to look critically at their ultraviolet spectra as well.

In the summer of 1976 I discovered that a significant class of organic materials that possess C=C double bonds in their structures have ultraviolet spectra that peak near the required wavelength. Moreover from the available laboratory measurements in the ultraviolet we could calculate that only 10% of the available carbon in space was needed in the form of this material to give a 2175Å band exactly of the observed strength.[8] This was a dramatic discovery and its publication in *Nature* carried the first exposition of the idea that the ultraviolet extinction band was due to complex organic molecules.

---

[8] F. Hoyle and N.C. Wickramasinghe, 1977. Identification of the λ2200Å interstellar absorption feature, *Nature*, 270, 323–324.

The year 1977 was a vintage year for my collaboration with Fred Hoyle, with no less than six papers being published in *Nature*. We were moving inexorably in the direction of astrobiology, possibly 20 years ahead of all our rivals and competitors. It should also be noted that we were remarkably successful in publishing most of our work in the so-called "high impact" journals up to this point. The campaign of outright censorship had not yet begun.

In the world at large, outside science, there were momentous events afoot: in Britain the Queen celebrated her Golden Jubilee, space sensationalism glutted Hollywood with movies like "Star Wars" and "Close Encounters of the Third Kind", almost as if the populace at large was getting ready for the arrival of ET in some form! In my native country, Sri Lanka, Junius Jayawardene (my father's school friend) becomes Prime Minister. The constitution of Sri Lanka is thereafter amended and Jayewardene becomes its first Executive President in 1978.

In January 1977 Fred Hoyle made his first long visit to Cardiff in his capacity as Honorary Professor. During this week-long visit, he stayed with us in Lisvane, as he would do on numerous subsequent occasions over the next decade and a half. It was always a pleasure to entertain Fred, and as the years went by, we became increasingly relaxed in his company. Priya is a charming and peerless hostess, and her efforts, particularly her cooking (she is an award-winning chef and cookery writer), were greatly appreciated by Fred. In fact she later came to joke with him about her contributions to man and science. Our friends, associates, and many others, were inevitably keen to meet Fred when he was in Cardiff, so places at our table were in high demand! We would invite small groups of guests to dinner, and Fred was charming company on such occasions, and his anecdotes were varied and surprising. I remember that once he nonchalantly related how, when he had finished the script for *A for Andromeda,* he toured the reparatory theatre in search of a "boyish young girl" to play the lead role. He was arrested by a performance of an unknown actress, Julie Christie, who agreed to play the part. By the second series, however, she had acquired such a degree of fame that she was too expensive to re-hire.

As far as our collaboration was concerned we now began to look at new ultraviolet spectra obtained for an extract of organic molecules from the Murchison meteorite. The data was supplied to us by A. Sakata in Tokyo, and we could immediately see here that the spectrum possessed an ultraviolet absorption feature near 2175Å, very similar to the observed interstellar extinction feature to which we have referred earlier. We submitted a short letter to *Nature* making this point which confirmed our earlier contention of an organic carrier for the 2175Å interstellar band.[9] We were slowly but surely beginning to ditch our graphite particle model for this feature in favour of organics — organics that turned out to be biochemicals in the fullness of time.

Our collaboration was now rapidly gathering momentum. We entered a phase involving a brisk exchange of telephone calls, letters and graphs between Cockley Moor and Lisvane. We had eventually stumbled upon one particular type of organic material that captured our interest more than any other. It was an infrared spectrum of cellulose. The laboratory spectrum of cotton cellulose over the 2.5–30 micrometre infrared waveband had shown, even at a cursory glance, most of the features required in order to explain astronomical spectra such as the Trapezium nebula. Cellulose is of course the main component of the cell walls of plants and is by far the most abundant terrestrially occurring biopolymer. It is logically at least the simplest biopolymer with a empirical formula $(H_2CO)_n$ just the same as for polyformaldehyde. It is a member of the most stable of a set of polymers known as the polysaccharides, which involves chains of various types of sugars, with a pyrolysis (heat destruction) temperature of 800K. The advantage of cellulose is that it could exist in regions of relatively high temperature of which the Trapezium nebula in Orion is an example.

My own instinct at this stage was to consider interstellar polysaccharides as being derived from life and being indicative of fully-fledged microbial life in the Galaxy. Fred, however, was still inclined to tread more carefully. He conceded the existence of polysaccharides or similar molecules in interstellar space, but still sought non-biological or abiotic processes for their formation.

The modelling of all these sources required a straightforward procedure known as a radiation transfer calculation which was a trivial job on the Cardiff University computer. But Fred insisted on checking everything himself and he performed all his calculations on a simple hand-held

---

[9]A. Sakata, F. Hoyle, N.C. Wickramasinghe, *et al*, 1977. Spectroscopic evidence for interstellar grain clumps in meteoritic inclusions, *Nature*, 266, 241.

programmable HP calculator that he carried everywhere. He maintained that being close to the logic of a calculation gave him a better insight into what was going on when it came to assessing the significance of the solutions that were obtained.

In parallel with our work on cellulose to model infrared spectra we also took another crack at the problem of understanding more fully the interstellar ultraviolet absorption. This was an extension of the work on the Murchison meteorite extract I referred to earlier. We argued that a class of double-ringed (bicyclic) aromatic compounds with the empirical formula $C_8H_6N_2$ (Quinozoline, Quinoxaline) provides an alternative explanation of the interstellar 2175Å absorption band.[10] Our paper contains the first suggestion in the literature of interstellar aromatic compounds and accords unequivocal priority to Fred Hoyle and myself for the idea of the existence of interstellar polyaromatic hydrocarbons — a presence that is now taken for granted with little or no credit being accorded to us. This work was published in an accessible and credible journal, and so I found a lack of referencing to it inexcusable by any standards of moral propriety.

To my mind the identification of polyaromatic hydrocarbons and cellulose-like polymers in interstellar space was tantamount to biology. It was the further pursuit of this line of reasoning that led to a surge of resistance from a highly conservative astronomical establishment. Seeing the direction in which we were moving, referees' reports of our later papers were becoming increasingly more disdainful. There was a point when my patience was tried by remarks that implied that the existence of organic polymers in space was impossible. They would all be destroyed by ultraviolet radiation, we were told, and indeed such comments were published by a number of distinguished astronomers including Carl Sagan and David Williams, a former President of the Royal Astronomical Society.

For example in a paper published in *Nature* W.W. Duley and D.A. Williams wrote thus:

"....We conclude that no spectroscopic evidence exists to support the contention that much of the interstellar dust consists of organic materials......." — *Nature*, 277, 4 January 1979.

---

[10] F. Hoyle and N.C. Wickramasinghe, 1977. Identification of the λ2200Å interstellar absorption feature, *Nature*, 270, 323–324.

Today, of course, everyone accepts without argument that complex organic molecules are everywhere in the galaxy.

At this point it should be put on record that two far-sighted individuals came to our rescue to ensure the rapid dissemination of our ongoing research. The first was Dr. C. W. L. Bevan, then Principal of University College, Cardiff, who chose to support our endeavours wholeheartedly in the belief that we were on the right track. The other was Zdenek Kopal who was the founding Editor and Editor-in-Chief of the journal *Astrophysics and Space Science,* and at the time Professor of Astronomy at Manchester. Kopal offered us the opportunity of publishing our ideas in his journal, whilst Bevan agreed to the funding by University College Cardiff of a "Blue Preprint Series". The now extinct University College Cardiff Press was also placed at our disposal for publishing preprints as well as monographs on subjects of our choice. In this way our channels of communication with the scientific community were not interrupted as the result of the unexpected turn that our researches were to eventually take.

It was at about this time that we began to feel the need for a dedicated laboratory facility to obtain spectroscopic data as and when we required them. Whilst still exploring the role of interstellar polysaccharides we realised somewhat belatedly that the Biochemistry Department of University College Cardiff headed by Ken Dodgson had on its lecturing staff a leading expert on polysaccharides, Tony Olavesen. We wasted no time to seek his help to measure for us the spectra of a large number of different polysaccharides under conditions we considered were appropriate for making comparisons with astronomy. This work soon gave us confirmation of our earlier conclusions that were based only on the published spectra of cotton cellulose, and led to another paper that was grudgingly published by the Journal *Nature.*[11]

In 1977, I received a letter from Dr. J. Brooks of the School of Chemistry at Bradford University including a spectrum of a complex organic biopolymer known as sporopollenin, which forms a major component of pollen and many spore walls. The spectrum had many of the features that were required

---

[11] F. Hoyle, A.H. Olavesen and N.C. Wickramasinghe, 1978. Identification of interstellar polysaccharides and related hydrocarbons, *Nature,* 271, 229–231.

to explain galactic infrared sources except for one fact. A 3.4 micron feature due to CH stretching was too prominent compared to the galactic sources we had seen. If, however, a thin layer of ice was condensed on particles comprised of sporopollenin, this difficulty would be rectified, thus making ice-coated bacterial spores a tenable model for interstellar dust. I promptly wrote to Fred Hoyle, who was in the United States at the time, soliciting his views and hopefully his support on this proposition. I soon followed this up with a draft of a joint paper for *Nature*. In a letter dated May 5 1977 written from Cornell University he wrote thus:

"Is the association of sporopollenin specific enough to support the final paragraph (speculating on the possibility of spores)? One might grant an interesting relation of the IR absorption obtained for the galactic centre or for galactic sources with the IR curve for sporopollenin, but the association may not require anything as complex biologically as sporopollenin itself...."

A few days later he reiterated the same point. In a letter dated May 9 also from Cornell he wrote:

"My feeling, however, is that we cannot invert the situation toward a conclusion favouring a particular complex biological structure like sporopollenin: Polymers→IR absorption O.K. But not IR absorption→Interstellar Biology"

A drastically toned down version of the first draft I sent to Fred (that included references to interstellar biology) eventually appeared in the journal *Nature*.[12]

Later that summer Fred Hoyle was back home in Cumbria. Still using his hand held HP computer he was calculating spectra on a large number of galactic infrared sources with opacity data for polysaccharides. The results, in terms of the closeness of fits, were most impressive. We first

---

[12] N.C. Wickramasinghe, F. Hoyle, J. Books and G. Shaw, 1977. Prebiotic polymers and infrared spectra of galactic sources, *Nature*, 269, 674–676.

issued our calculations as a preprint in our "Cardiff Blue Preprint Series", and later in two papers — a short version in *Nature*[13] and a fuller version in *Astrophysics and Space Science.*[14]

In July of this year, Phil Solomon, one of the pioneers in the detection of interstellar molecules using millimetre waves, visited me in Cardiff. He met with Fred both in Cardiff, and also in Mid-Wales for a workshop on Giant Molecular Clouds in the Galaxy that I had organised. The venue for the symposium was Gregynog Hall, a Conference Centre near the city of Newtown belonging to the University of Wales. With links dating back to the 12th century, Gregynog Hall is an imposing 19th century manor house set amidst 750 acres of gardens, woodland and farmland. It was bequeathed to the University in the 1960s by the Davies sisters, who had been avid art collectors, as well as generous patrons of the arts. Fred and Barbara Hoyle, along with their two grandchildren, rented one of the many cottages in the grounds. Priya and I, and our two young children, did the same. The kids were all of a similar age and enjoyed having free rein to tear around the vast grounds together. The conference itself brought together a small group of scientists, and was hailed as a success. Fred and I had ample time between sessions to discuss our strategy for further development our ideas. It is worth noting that the series of Gregynog astrophysics workshops I started after setting up the new Department of Applied Mathematics and Astronomy acquired a measure of international fame and attracted several scientific luminaries. During the period 1976–1979 our visitors included William Fowler, Cyril Ponnamperuma and Donald Clayton — all attending the Gregynog workshops.

In September 1977, we were abroad again, visiting the Astronomy Department of the University of Western Ontario, Canada. Fred was on his way to Caltech. Perhaps in need of a respite from the intensity of the life-in-space argument, Fred and I started pouring over a volume entitled "Cretaceous-Tertiary Extinctions and Possible Terrestrial and Extraterrestrial

---

[13] F. Hoyle and N.C. Wickramasinghe, 1977. Polysaccharides and infrared spectra of galactic sources, *Nature*, 268, 610–612.

[14] F. Hoyle and N.C. Wickramasinghe, 1978. Calculation of infrared fluxes from galactic sources for a polysaccharide grain model, *Astrophys. Sp. Sci.*, 53, 489–505.

Causes" published in the previous year by the Canadian Museum of Natural History in Ottawa. This led to a diversion from our obsessive interest in organic grains.

Some 65 million years ago the dinosaurs and indeed all animals with body weights above 25 kg suddenly became extinct. We argued that this could be due to the interaction of the Earth with a cloud of porous cometary dust derived from the extended coma of a comet. The Earth's stratosphere will be dusted over in a way that two thirds of the light and incident energy from the sun will be blocked for several years whilst still permitting infrared radiation to leak out. The result would be semi-darkness for a decade, leading to the withering of foliage in trees and causing a severe interruption of food chains. Herbivorous creatures including dinosaurs would soon become extinct, and so would carnivores that feed on the herbivores. With rivers still continuing to run and some lakes remaining unfrozen, fresh water organisms would survive — their food chains depending on decaying vegetable matter, would take longer to disrupt than marine organisms dependent on phytoplankton. The seeds and nuts of land plants would also survive and small animals, including small mammals, living on nuts and seeds would also survive the dark and desolate years. We humans owe our descent through this ecological crisis to the survival of these small mammals. All these ideas were published in the form of an article in *Astrophysics and Space Science*.[15]

Our ideas on the Cretaceous-Tertiary extinctions were similar (though not identical) to those of Alvarez and Alvarez that were published approximately two years after ours, and which have come to be more or less generally accepted. A direct hit by a comet seems to have occurred 65 million years ago, and the resulting crater has been discovered in the seabed of the Yuccatan Peninsula.

In addition to causing extinctions of species we also argued in our paper that cometary dusting over a more protracted period could trigger the onset of ice ages. This process was explored by us in greater detail in the late 1990's.

---

[15] F. Hoyle and N.C. Wickramasinghe, 1978. Comets, ice ages and ecological catastrophes, *Astrophys. Sp. Sci.*, 53, 523–526.

CHAPTER 12

DISEASES FROM SPACE

Serendip is an ancient name for the island of Sri Lanka and was in use from the 4th Century AD. In the fairy-tale of Horace Walpole (1717–1797), "Three Princes of Serendip", the heroes keep making delightful discoveries of things that they were not in quest of. This, according to the Oxford English Dictionary is the origin of the English word serendipity.

It is perhaps not a surprise that serendipitous events played a part in my researches at crucial stages of its development. This was certainly so for events that were to lead to a particular diversion that was to engage my attention almost obsessively for a full three decades. It was a curious tale that started with sniffles in the late summer of 1977, just prior to my trip to Canada. I had succumbed to an unseasonal bout of flu-like illness and this happened at a time when Hoyle and I were in a phase of brisk telephone exchanges over matters relating to the origin of life in comets. I was suddenly reminded of my mother's admonition in my childhood: "Don't go out in the rain or in evening mists or you'll get ill!" A similar belief is, of course, widely prevalent in the West. And in many cultures throughout the world comets have also been thought of as harbingers of pestilence and death. Could time-hallowed beliefs possibly have their basis in hard fact?

I telephoned Fred Hoyle in Cockley Moor, on a depressingly grey afternoon in 1977, with perhaps the most preposterous proposition I had ever made. Could the old wive's tales of diseases being connected with rain have possibly been right? Could viruses be present in comets, and could cometary viruses entering the Earth cause disease? Fred was caught unawares and I was greeted by a long silence at the other end, before he finally said he would think about it. I was indeed extremely surprised when he phoned back within hours of my call agreeing that this might be so. Fred was reminded of conversations he had had some years earlier with the Australian physicist E.G. (Taffy) Bowen. Bowen had discovered an amazing connection between freezing nuclei in rain clouds and the incidence of extraterrestrial particles. What Taffy Bowen had showed was that there was a link between the frequency of freezing nuclei in tropospheric clouds and the occurrence of meteor showers. Meteor showers occur at regular times in the year whenever the Earth in its orbit crosses the trails of debris evaporated from short-period comets. If bacteria and viruses come in with cometary meteor showers they could, if they survive entry, act as freezing nuclei for rain. Raindrops laden with bacteria and viruses then become a distinct possibility. By analysing all the available data Bowen had reported in a paper in *Nature* (**177**, 1121, 1956) that as dust from meteor streams falls into the troposphere heavy falls of rain could be expected. This is found to happen some 30 days after the meteor dust first entered the very high atmosphere. It was discovered only much later that bacterial and fungal spores could indeed lower the freezing temperature of water and act as condensation nuclei for rain.

Because Fred was still hesitant in 1977 to accept the possibility of bacterial grains in interstellar space, his instant conversion to the idea of disease-causing bacteria and viruses coming from comets somewhat surprised me. The idea that now had to be tested was that comets carry bacteria and viruses and that their continued interaction with the Earth could bring new pathogens that could affect plants and animals from time to time. By observing the pattern of appearance of diseases in the past we soon discovered that a strong *prima facie* case existed in support of our contention. The rather sudden appearance in the literature of references to particular diseases is significant in that it probably points to times of specific "invasions". Thus the first clear description of a disease resembling influenza is early in the 17th century AD. The common cold has no mention until about the 15th century AD. Descriptions of small pox and measles do not appear in a clearly recognisable form until about the 9th century AD. Furthermore, certain early plagues such as the plague of Athens of 429 BC, which is vividly detailed by the Greek historian Thucydides, do not seem to have an easily recognisable modern counterpart.

We noticed that epidemics and pandemics of fresh diseases, both in historical times as well as more recently have almost without exception appeared suddenly and spread with phenomenal swiftness. The influenza pandemics of 1889–1890 and 1918–1919 both swept over vast areas of the globe in a matter of weeks. Such swiftness of spread, particularly in days prior to air travel, is difficult to understand if infection can pass only from person to person. Rather it is strongly suggestive of an extraterrestrial invasion over a global scale. We argued now that it is the primary cometary dust infection that is the most lethal, and that secondary person-to-person transmissions have a progressively reduced virulence, so resulting in a diminishing incidence of disease over a limited timescale.

On this picture, the pattern of incidence and propagation of any particular invasion is a somewhat complex matter. It depends, amongst other factors, upon the sizes of incoming micrometeoroids, the local physical characteristics of the atmosphere and on the distribution of global air currents. We might expect certain latitude belts on the Earth to be comparatively disease free, whilst others might be more prone to receiving space-borne pathogens. Also, depending on sizes of particles amongst other factors, some epidemics may be geographically localised while others may be global. In all cases any new epidemic must occur suddenly — when the Earth crosses the trail of infected cometary particles.

We became fully convinced that these ideas had to be substantively right. We both made a great effort to learn as much as we could about virology and infectious diseases, from text books as well as by talking to our medical colleagues at the University Hospital, particularly John Watkins, Professor of Medical Microbiology and Robert Mahler, Professor of Medicine. Fred Hoyle's visits to Cardiff were now taken up with locating the right virologists, bacteriologists and historians who could enlighten us with facts about diseases past and present. We also made several visits together to the Central Virus Reference Laboratory in Colindale, London, and one memorable visit to meet Sir Christopher Andrewes (a virologist who played a key role in isolating cold-type viruses) at the Common Cold Research Centre in Salisbury plains. Here we discovered that all attempts to infect volunteers with a common cold virus under controlled, epidemic-like conditions had been, up to that time, a failure.

We soon began to take a particular interest in one disease — influenza — because we discovered there were many puzzling aspects connected with its epidemiology (epidemic behaviour). Here was a disease that appeared to indicate the incidence of a virus or at least a trigger for it from the atmosphere. A distinguished epidemiologist Charles Creighton maintained as late as the final decade of the 19th century that influenza is not a transmissible disease. In his book "History of Epidemics in Britain" (Cambridge University Press, 1891) he discusses the influenza epidemics of 1833, 1837, and 1847, in which medical opinion held that populations living over considerable areas are affected almost simultaneously. Such evidence suggested to Creighton a "miasma" descending over the land rather than a disease which must spread itself from person to person. If one substitutes for "miasma" the phrase "viral invasion from space" it is a similar position to that which we arrived at in 1977. Creighton's hegemony, however, was short lived, and by the end of the 19th century, the concept of infectious disease caused by microorganisms had firmly taken root.

Strictly, the microbiological concept requires only that victims of the disease should acquire the virus from outside of themselves, which of course they would do if the infection came from the atmosphere. But such an idea seemed so much less plausible to scientific opinion than the concept of person-to-person transmission — this was not even considered as a hypothesis to be tested. It became an axiom.

We became convinced that the transmission hypothesis could be tested if a new pandemic strain arose. Particles of viral size added to the Earth are stopped in the atmosphere at a height of about 30 km. Vertical descent of particles through the stratosphere and into the troposphere occurs mainly as a result of winter downdrafts that occur six-months apart in the two hemispheres. This phenomenon offers a ready explanation of the fact that influenza and other respiratory viral diseases are distinctly seasonal in character.

It is commonly found that lingering mists in the winter season usher in a wave of flu-like disease. Since, as I have already said, bacteria and viruses can act as condensation nuclei around which water droplets form, this coincidence is not entirely unexpected. In situations where rain falls as large drops there is not much chance of direct inhalation of nucleating viruses, whereas misty weather provides the incoming virus with the best opportunity to become dispersed in aerosol form in a way that can easily be inhaled near ground level.

Among earlier evidence that pointed in this direction, the observations of Professor Magrassi in 1948 are worthy of note. The world-wide epidemic of 1948 apparently first appeared in Sardinia. Magrassi,[16] commenting on this epidemic, wrote thus:

"We were able to verify the appearance of influenza in shepherds who were living for a long time alone, in open country, far from any inhabited centre; this occurred almost contemporaneously with the appearance of influenza in the nearest inhabited centres..."

One of the most striking features in this whole story is that the technology of human travel has had no effect whatsoever on the way that influenza spreads. If influenza is indeed spread by contact between people, one would expect the advent of air travel to have heralded great changes in the way the disease spreads across the world. Yet the spread of influenza in 1918, before air travel, was no faster, and no different from its spread in more recent times.

Probably the most disastrous influenza pandemic in recent history occurred in 1918–1919 and caused about 30 million deaths. After studying all the available information about the spread of influenza during this epidemic Dr. Louis Weinstein wrote in the May 1976 issue of the *New England Journal of Medicine*:

"Although person-to-person spread occurred in local areas, the disease appeared on the same day in widely separated parts of the world on the one hand, but on the other took days to weeks to spread relatively short distances. It was detected in Boston and Bombay on the same day, but took three weeks before it reached New York City, despite the fact that there was considerable travel between the two cities. It was present for the first time at Joliet in the State of Illinois four weeks after it was first detected in Chicago, the distance between those areas being only 38 miles..."

---

[16] F. Magrassi, 1949. *Minerva Med. Torino*, **40**, 565.

As we were pondering on such matters serendipity intervened once again with a remarkable circumstance. The new pandemic we had talked about had become a reality. In November 1977 an outbreak of a strain of flu that had not been present in the human population for 20 years, was reported in the Far East region of the old Soviet Union. By the end of December the first cases were reported in Britain and by January the disease was rampant, most noticeably it was raging through the schools of England and Wales.

Our ideas on the emergence of life on comets and of a possible connection with plagues and pestilences had now advanced to a stage when I felt that Fred Hoyle should try them out on an academic audience. He obligingly agreed to deliver a lecture entitled "Diseases from Outer Space" at Cardiff on January 18, 1978. Needless to say the lecture was a sell-out. There was only standing room in the main auditorium of Cardiff's largest lecture theatre at the time, the Sherman Theatre. His lecture was greatly appreciated even though it stimulated controversy and a degree of hostility as well. The so-called Red Flu was around us everywhere, as was evident even in the coughing that was heard through the lecture. During the days he spent in Cardiff, Fred Hoyle and I developed a strategy to investigate this epidemic.

We saw this as an ideal opportunity for testing person-to-person transmission. School children under the age of 20 had not been exposed to the new virus in their lifetime, and so were all equally susceptible. We had the idea of using the school population as detectors of the new virus, rather in the same way that physicists use amplifying detectors to observe small fluxes of incident cosmic rays. Our first "experiment" was confined to schools, including boarding schools in South Wales and the South West region of England. We began with analyses of school absenteeism. For this purpose Priya and I did a mammoth circuit of schools within 30 miles of our house, and examined their attendance records. Our object was to determine, for each individual school, the time dependence of absenteeism due to influenza during the epidemic, and also the overall attack rates that could be inferred. What surprised us most was the huge range in the attack rates, essentially 0% to over 80%, and this was determined only by the location of the school, indicating that the incidence of the virus was patchy on distance scales of a few kilometres or less. Data that Priya and

I had collected for boarders in Howells School Llandaff, Cardiff and Atlantic College at St Donats gave a taste of more to come. There was clearly an effect connected with the houses where the children slept, some had high attack rates, others very low. Already, person-to-person transmission was beginning to look unlikely.

We published our preliminary findings in *New Scientist* (September 28, 1978) and then started a more ambitious exploration of the school data throughout England and Wales. We circulated a questionnaire to all privately supported secondary schools seeking the following information:

(1) Over-all attack rates amongst boarders and day pupils (separate data) during the recent influenza epidemic.
(2) Day-to-day attack patterns as they are reflected in classroom absence and/or admissions to school sick bays.
(3) The date of the peak of epidemic experienced in each school.

The results of our bigger survey only confirmed and strengthened those we had from the local schools. It is commonly stated that school boarders, members of the armed services in stations, and other closed communities, are highly susceptible to epidemic outbreaks of influenza. Replies to our questionnaire showed clearly that as far as school boarders are considered this is a myth. Our sample involved a total of more than 20,000 pupils with a total number of victims of some 8,800. The distribution of attack rates in the schools showed that only three schools out of more than a hundred had the very high attack rates that have been claimed to be the norm.

If the virus responsible for the 8,800 cases were passed from pupil to pupil, much more uniformity of behaviour would have been expected. We found evidence for great diversity, with a hint that the attack rate experienced by a particular school (or house within a school) depended on where it was located in relation to a general infall pattern of the virus. The details of this infall pattern are determined by local meteorological factors. The infall clearly displayed patchiness over a scale of tens of kilometres, the typical separation between the schools.

One particular school in our survey, Eton College, merits special attention. There were 1248 pupils distributed in a number of boarding houses and the total number of cases across the whole school was 441. The actual distribution of cases by house showed enormous heterogeneity. College house with a total population of 70 had only one case, compared with the expected value of 25 on the assumption of random distribution, in a person-to-person infection model. Here again we saw heterogeneity, but now on the scale of hundreds of metres. This entire distribution would be expected once in 10^{-16} trials on the basis of person-to-person transmission. Clearly, if one looks objectively at all the facts, flu cannot be "catching" from person to person as our present-day scientific culture would have us to believe.

Further evidence against the standard dogmas of influenza transmission came from a study of influenza in Japan. The Japanese data was supplied by my friend Shin Yabushita, a contemporary of mine at Cambridge. Japan is remarkable in that several large areas have population densities in excess of 2000 per square km, whereas others have well under 200 per square km. Standards of monitoring and reporting disease are also uniformly efficient throughout most of Japan. The data showed extreme variability of attack rates between adjacent prefectures. Here again the patchiness of viral in-fall is over a distance scale of tens of kilometres, exactly as in the case of the Welsh and English schools we already saw.

We have already noted that the descent of the virus from the stratosphere to ground level depends on global circulation patterns of the atmosphere. This fact accounts for the otherwise mysterious phenomenon that epidemics of flu occur with a distinct seasonality in widely separated parts of the world.

Scientists who feel uncomfortable with the logical inferences drawn from a theory such as this often seek eye-catching one-line disproofs. For the case of influenza from space the often-stated disproof is that: "Viruses are host-specific, and so must have evolved in close proximity to the terrestrial species that they attack." In other words the critic says: "How could the incoming virus know ahead of its arrival the nature of the highly evolved and specialised hosts that it may encounter". Our answer is simple: "The virus could not of course anticipate us, but we, the host species, could anticipate the virus since we must have had a long and continuing exposure to viruses of a similar kind." It is also well known that viruses can on occasion add onto our genes and so viral DNA sequences serve as an invaluable store of evolutionary potential. Our genomes would be "made up" of cometary viruses according to this point of view. If the influenza virus, or one that is similar to it, forms part of our genetic heritage, then the so-called host specificity, or the apparent human-virus connection is instantly and elegantly explained. This point of view has recently found support in results obtained from sequencing of the human genome. It is now known that a large fraction of the human genome is indeed comprised of viral sequences.

According to our point of view, reservoirs of the causative agent for influenza are periodically resupplied at the very top of the Earth's atmosphere. Small particles, the sizes of viruses or smaller, tend to remain suspended high up in this region for long periods unless they are pulled down into the lower atmosphere. In high latitude countries, such breakthrough processes, where the upper and lower air becomes mixed, are seasonal and occur during the winter months. Thus a typical influenza season in a European country would occur between December and March. Frontal conditions with high wind, snow and rain effectively pull down viral pathogens close to ground level. The complex turbulence patterns of the lower air ultimately control the details of the attack at ground level, and determine why people at one place and at one time succumb, and why those in other places and at other times do not.

We also discovered in the literature that ozone measurements can be used to trace the mass movements of air in the stratosphere. Such measurements show a winter downdraft that is strongest over the latitude range from 40° to 60°. Taking advantage of this annual downdraft, individual viral particles incident on the atmosphere from space would therefore reach ground-level generally in temperate latitudes. Such locations on the globe would naturally emerge as places where upper respiratory infections are likely to be most prevalent, on the supposition of course that the Earth is smooth. The exceptionally high mountains of the Himalayas, rearing up through most of the height range to the stratosphere, introduce a large perturbation on the smooth condition, which may be expected to affect adversely this particular region of the Earth, especially regions lying downwind of the Himalayas, particularly China and South East Asia. In effect, the Himalayas are so high that they could act as a drain plug for most of the viruses incident on the atmosphere at latitude ~30°N, the large population of China being inundated by this drainage effect, making China the quickest and worst affected region of the Earth. This could explain why new respiratory viruses such as SARS and new influenza viruses often make their first appearance in China. Concomitantly, other parts of the Earth at ~30°N should be largely free of viral particles, unless it happens that such particles are incident as components within larger particles which fall fast under gravity.

A direct demonstration that the general winter downdraft in the stratosphere occurs strongly over the latitude range 40° to 60° was given by M.I. Kalkstein (Science, 137, 645, 1962) in the last of the series of atmospheric nuclear tests carried out in the middle of the 20th century. A radioactive tracer, Rh-102, was introduced into the atmosphere at a height above 100 km and the fall-out of the tracer was then measured year by year through airplane and balloon flights at altitudes ~20 km. The tracer was found to take about a decade to clear itself through repeated winter downdrafts, and this happened mostly over the latitude belt 40–60 degrees.

The observed fall out patterns of a radioactive tracer agreed closely with the well-known winter season of the viruses responsible for the majority of upper respiratory infections in temperate latitudes. The time of a decade or so that was taken to clear the tracer from the stratosphere also coincides with the average time of prevalence of any new influenza subtypes after it is first introduced. With all these considerations well in place on influenza as well as other epidemic diseases we began to write our second book on the subject *"Diseases from Space"*, published in 1979, again by J. M. Dent.

CHAPTER 13

GALAXIES STREWN WITH MICRO-ORGANISMS

The year 1977 marked several historic events. Queen Elizabeth II celebrated her Silver Jubilee. Jimmy Carter was elected President of the United States. More relevant to my own story President Junius Jayewardene became Executive President of the Republic of Sri Lanka. In science the bacterium causing epidemics of Legionnaire's disease was finally isolated.

By 1977 I had come to the conclusion that the chemical composition of interstellar dust (judged by spectral features) was unequivocally organic, and the best types of organics that matched all the data were similar to biochemicals. However, there were many unresolved problems to be tackled. Over the visible wavelength range from 7000 to 3000Å, the extinction of starlight (dimming ratio on a logarithmic scale) was approximately inversely proportional to the wavelength, and this was the case in whatever direction one looked. Such an invariance of behaviour (as observed first by Kashi Nandy) was difficult to reconcile with the grain models that were being discussed in orthodox astronomy involving mixtures of inorganic grains. In all such non-biological grain models the visual part of the extinction curve of starlight had to come mostly from the dielectric (nonabsorbing) component of the mixture, and to get the correct shape of this curve, grain radii had to be fairly precisely fixed. In other words, the solutions obtained so far in all these cases were highly parameter-sensitive, and therefore not very satisfactory.

In 1977 I discovered that one way to solve this problem is to have hollow organic particles instead of fully solid grains. I got to work on discussing ways in which bacteria would freeze-dry in space leading to

119

the production of vacuum cavities within them — thus producing exactly the required type of hollow grain. I started with a vegetative bacterial cell with the following typical constitution: Organic material 20%, bound water 20%, free water 60%. Freeze-drying in a vacuum, such as which exists in space, would maintain the cell wall intact and also retain the interior organic content and bound water, while the free water will escape and lead to the production of vacuum cavities to almost exactly what was required for the model to work.

The next thing we needed to know was the size-distribution that would be representative for actual bacteria and for this *Bergey's Manual of Determinative Bacteriology* was consulted. The best data we could find related to spore-forming bacteria, which gave the count of species in various ranges of sizes in the form of a histogram. By taking freeze-dried interstellar bacteria to have the particular size distribution that was appropriate for terrestrial spore forming bacteria, we had a situation totally different from anything we had experienced previously. For now there were no parameters at all to be fitted. The extinction behaviour of the entire ensemble of grains became immediately amenable to calculation using the computer program I had developed and used over many years. The result was staggering: we had discovered a perfect fit to the average interstellar extinction over the visual waveband with just the one assumption — that interstellar grains were freeze-dried bacteria.

I believe it was the extraordinary nature of this fit, coming as it did after two decades of failure, that won Fred Hoyle over to the case for bacterial grains. And then there was no stopping or turning back. For unravelling the composition of interstellar dust we had progressed cautiously and in slow stages through a sequence of options: graphite, organic polymers, and then to complex biopolymers such as the polysaccharides. These organic polymeric particles that were in evidence everywhere in the galaxy had an average size comparable to a bacterium, and an average refractive index appropriate for a freeze-dried bacterium. Good fits to infrared, visual and infrared data were possible on the single assumption of bacteria-like particles. Could all this be somehow explained without invoking biology? Of course this question has to be explored.

After weeks of fumbling through a sequence of ideas, all of which were proving to be woefully inadequate, we alighted on the most promising, if not outrageous hypothesis. In the gigantic clouds of interstellar dust could we be witnessing no less than the dissemination of biology? Could the interstellar medium be choc-a-bloc, not simply with the building blocks of life arranged into prebiotic molecules or even protocells, but with the end products of the living process? And this would then be required to happen on an unimaginably vast scale. At the end of a long run of frenzied telephone calls between Cardiff and Cockley Moor, Fred Hoyle and I decided that was it! Interstellar grains must surely be bacteria — albeit freeze-dried, perhaps mostly dead! At the very least this was a hypothesis that had to be fully explored.

Nobody takes lightly the prospect of walking into exile, albeit scientific exile, and moreover one that is self-imposed. Yet by the Spring of 1977 it appeared that we had no option but to do so, carrying the heavy burden of not one but two scientific heresies. The heresy of diseases from space and now the heresy of microbial life in interstellar space. No sooner we obtained this solution we wrote up our results, distributed them to colleagues in the form of a Blue Cardiff Preprint at the beginning of April 1979, and at the same time sent off a technical paper for publication in *Astrophysics and Space Science.*[17]

The immense power of bacterial replication is worth careful note at this point. Given appropriate conditions for replication, a typical doubling time for bacteria would be two to three hours. Continuing to supply nutrients, a single initial bacterium would generate some 2^{40} offspring in four days, yielding a culture with the size of a cube of sugar. Continuing for a further four days and the culture, now containing 2^{80} bacteria, would have the size of a village pond. Another four days and the resulting 2^{120} bacteria would have the scale of the Pacific Ocean. Yet another four days and the 2^{160} bacteria would be comparable in mass to a molecular cloud like the Orion Nebula. And four days more still for a total time, since the beginning, of 20 days, and the bacterial mass would be that of a million galaxies. No abiotic process remotely matches this replication power of a biological template. Once the immense quantity of organic material in the interstellar material has been appreciated, a biological origin for it becomes a necessary conclusion.

---

[17] F. Hoyle and N.C. Wickramasinghe, 1979. On the nature of interstellar grains, *Astrophys. Sp.Sci.*, 66, 77–90.

But where are astronomical locations where conditions for replication of bacteria can be found? Certainly not in the cold depths of space, where microbes could merely remain in a freeze-dried dormant state. Planets like the Earth would provide too small a total mass of carbonaceous material to make any cosmic impact. It is therefore to comets we turned, arguing that comets are the sources of biological particles in interstellar clouds. An individual comet is a rather insubstantial object. But our solar system possesses so many of them, perhaps more than a hundred billion of them, that in total mass they equal the combined masses of the outer planets Uranus and Neptune, about 10^{29} grams. If all the dwarf stars in our galaxy are similarly endowed with comets, then the total mass of all the comets in our galaxy, with its 10^{11} dwarf stars, turns out to be some 10^{40} grams, which is just the amount of all the interstellar organic particles that are present in the dust clouds within the galaxy.

How would microorganisms be generated within comets, and then how could they get out of comets? We know as a matter of fact that comets do eject organic particles, typically at a rate of a million or more tons a day when they visit the inner regions of the solar system. We argued that comets when they are formed incorporate interstellar bacterial particles, from which only a fraction 10^{-22} needs to retain viability for a regeneration process to operate. For at least a million years, at the time of their origin, comets have liquid cores due to radioactive heat sources such as ^{26}Al which are also incorporated within them. Within a very brief period as described above, sequential doublings of viable microorganisms would lead to an entire cometary core being converted into biomaterial. When the comets re-freeze this amplified microbial material is also frozen in, only to be released when they become periodically warmed up in the inner solar system. Some of this bacterial matter may reach the inner planets, which they can seed with life; some of it is expelled back into interstellar space.

Our point of view requires that bacteria must be space-hardy which recent research has shown is the case. On the whole, microbiological research of the past 20 years has shown that bacteria and other microorganisms are indeed remarkably space-hardy. Microorganisms known as thermophiles and hyperthermophiles are present at temperatures above boiling point in oceanic thermal vents. Entire ecologies of microorganisms are present in the frozen wastes of the Antarctic ices. A formidable total mass of microbes exists in the depths of the Earth's crust, some eight kilometres below the surface, greater than the biomass at the surface. A species of phototropic sulfur bacterium has been recently recovered from the Black Sea that can perform photosynthesis at exceedingly low light levels, approaching near total darkness.[18] There are bacteria (e.g. *Deinococcus radiodurans*) that thrive in the cores of nuclear reactors. Such bacteria perform the amazing feat of using an enzyme system to repair DNA damage, in cases where it is estimated that the DNA experienced as many as a million breaks in its helical structure. And the list goes on. There is scarcely any set of conditions prevailing on Earth, no matter how extreme, that is incapable of harbouring some type of microbial life.

In the years 1979 and 1980 when these crucial steps were taken in our collaboration, the world outside was recording dramatic events. The Shah of Iran was driven into exile; for home rule in Scotland and Wales, the Scots say yes, the Welsh say no; Margaret Thatcher becomes Britain's first woman Prime Minister. More relevant to our story the US spacecraft Voyager I returns the most dramatic pictures ever of Jupiter, Saturn and Uranus. Rings are discovered around both Jupiter and Uranus, leading us to speculate whether these contained bacterial particles.

On October 29th–31st 1980, a conference with the title "Comets and the Origin of Life" was organized by Cyril Ponnamperuma at the University of Maryland. Ponnamperuma, a Sri Lankan-born chemist always had his feet firmly planted in the opposite camp from us. He had made important laboratory studies on the abiotic synthesis of organic molecules, including sugars and nucleotide bases under conditions similar to those used in the famous Urey–Miller experiment of the 1950's. All such experiments are of course a very far cry from the generation of life, but they are often presented as being a significant step in that direction. Ponnamperuma, at the time was Director of the Laboratory of Chemical Evolution at the University of Maryland, and clearly was an opponent of the ideas being discussed by Fred Hoyle and myself. I was invited by him to present a joint paper with Hoyle entitled "Comets — a vehicle for panspermia", an invitation that I accepted with some trepidation.

---

[18] J. Overmann, H. Cyoionka & N. Pfennig, 1992. *Limnol.Oceanogr.*, 33 (1), 150–155.

Cyril Ponnamperuma had rounded up just about all of our potential adversaries, including J. Oro and of course J. Mayo Greenberg. Almost everyone at the meeting delivered a polemic denying of any possibility of comets carrying life, or even life's building blocks in some instances. Greenberg made his first coherent claim to rival our 1974 ideas of polymeric grains with his own title "Chemical Evolution of Interstellar Dust — A Source of Prebiotic Molecules". He was unfortunately a few years too late to steal any priority from Fred Hoyle or myself. The conference volume, including our paper, was published by D. Reidel Co. in 1981 (*Comets and the Origin of Life* — ed. C. Ponnamperuma). Nothing that I heard at this meeting made me weaken my resolve to continue our own search for an origin of life on a grand cosmic scale. Facts were moving swiftly in our direction and the die-hard opposition understandably seemed to die hard.

With the emergence of Cardiff as a new centre for astronomy, the Royal Astronomical Society asked me to arrange their annual out of town meeting for 1980 in Cardiff. The then RAS President, Professor M.J. Seaton of University College London also requested Fred Hoyle to give an evening public lecture on this occasion. On April 15, 1980 Fred Hoyle delivered his lecture with the title "The Relation of Biology to Astronomy" at University College Cardiff, in which he presented an eloquent exposition of our position on the nature of interstellar grains. His frontal assault on conventional theories of biological evolution on the Earth did not win him much support, but the case against such theories was outspokenly presented. For instance:

"What may be the biggest biological myth of all holds that evolution by natural selection explains the origin of the phyla, classes and orders of plants and animals. There are certainly plenty of examples of minor evolutionary changes caused by natural selection, and on the evidence of these minor changes the major changes are assumed to be similarly caused. The assumption became dogma, and then in many people's eyes the dogma became fact....."

This was a statement of intent that we were soon to take on the entire biological establishment in re-evaluating the evolution of life in a closed box setting, and opening up the process to the wider universe. This was to happen in the months that followed leading eventually to the publication of our book entitled *Evolution from Space*.

I suspect that the majority of the audience in Cardiff, who were astronomers, did not take much interest in biology at this stage even though it was the dominion of astronomy that was being re-evaluated and enlarged. Thus Fred Hoyle went on:

"Astronomers have become accustomed to thinking of the external Universe in the words of Macbeth, as being "full of sound and fury, signifying nothing". Can we seriously believe that anything as subtle as biology could have gained a toehold in a world signifying nothing? I pondered this question for a long time before arriving at a strange answer to it. If the astronomer's world of fury is really in control, then the prospects for biology would be poor. But what if it is really biology which controls the astronomer's world?"

Unwittingly perhaps Fred laid the foundations for the modern discipline of astrobiology, a subject that is becoming increasingly popular these days, with his concluding remarks on April 15, 1980:

"Microbiology may be said to have had its beginnings in the nineteen-forties. A new world of the most astonishing complexity began then to be revealed. In retrospect I find it remarkable that microbiologists did not at once recognise that the world into which they had penetrated had of necessity to be of a cosmic order. I suspect that the cosmic quality of microbiology will seem as obvious to future generations as the Sun being the centre of our solar system seems obvious to the present generation."

In the days immediately following the RAS meeting Fred Hoyle stayed with us in Cardiff as usual. We took this opportunity to identify many of the loose ends of our theory that had to be dealt with. The immediate question now was: what further checks were there to be made for the thesis of interstellar bacteria? Precise predictions of the behaviour of a bacterial model at infrared wavelengths should be made, and these might then be looked for in astronomy. But the required experimental work was preferable to be done *ahead* of astronomical observations being made.

Here was our next instance of the intervention of serendipity. My brother D. T. Wickramasinghe, (Dayal), Professor of Mathematics at the Australian National University in Canberra, was also an astronomer and frequently used the 3.9 metre Anglo-Australian Telescope, that happened to be equipped with just the right instruments to look for a signature of interstellar bacteria.

Shortly after the release of our April 1979 preprint on the scattering properties of bacteria, Dayal visited Cardiff to spend some time with our family. Dayal's visit happened to coincide with a time when Fred Hoyle was also in Cardiff. We naturally got talking about matters relating to interstellar bacteria. Dayal asked: "What do you think can be done at the telescope to prove or disprove your theory?" to which we promptly replied that he could use the infrared spectrometers on the AAT to look at infrared sources near the wavelength of 3.4 micrometres in greater detail than before. A very long path length through the galaxy was needed to have any hope of detecting such an effect unambiguously. The longest feasible path length through interstellar dust that existed within our own galaxy was defined by the distance from the Earth to the centre of the Galaxy. There were several sources of infrared radiation located near the galactic centre that could serve as search lights for interstellar bacteria. Dayal was doubtful that he would be allocated observing time if he applied for such time specifically to do this project. The general consensus then was that life in space could not be regarded as respectable science! Dayal overcame this difficulty, however. Although honesty is the best policy it often pays handsomely to be economical with the truth in a world of dubious morality. The deceit involved applying for telescope time to do a quite different project, and then illicitly using part of the time to look for the signature of organic matter.

In February and April 1980, Dayal, collaborating with D. A. Allen, obtained the first spectra of a source known as GC-IRS7 which showed a broad absorption feature centred at about 3.4 micrometres.[19] When Dayal's published spectrum was examined we found that it agreed in a general way with the spectrum of a bacterium that we had found in the published literature. But at this time neither the wavelength definition of the

---

[19]Wickramasinghe, D.T. and Allen, D.A., 1980. *Nature*, 287, 518–519.

astronomical spectrum nor the laboratory bacterial spectrum was good enough to make a strong case for interstellar bacteria. However, even from this early observation we were able to check that the overwhelming bulk of interstellar dust must have a complex organic composition.

It was precisely at this moment that Shirwan Al-Mufti, a practical man, the son of an Iraqi Army General, approached me to become a research student at Cardiff. Here was our chance to get the required laboratory work done. We approached Tony Olavesen at the Biochemistry Department at Cardiff and arranged for Al-Mufti to be given bench space and laboratory facilities in that Department to undertake spectroscopic studies of biological samples. The purchase of a modest amount of equipment that was needed was immediately authorised by Principal Bevan, and our experimental project got under way. Al-Mufti set about executing this task with military-style efficiency, and the experiments began to yield important results in the first few months of 1981.

Al-Mufti's experiments involved desiccating bacteria, such as the common organism *E. Coli*, in an oven in the absence of air and measuring, as accurately as possible, the manner in which light at infrared wavelengths is absorbed. The normal technique for doing such a measurement involved embedding the bacteria in discs of compressed potassium bromide and shining a beam of infrared light through them. The standard techniques had to be adapted only slightly. There was the need to match the interstellar environment which involved desiccation, and the spectrometer that was used had to be calibrated with greater care than a chemist would normally exercise. When all this was done it turned out that a highly specific absorption pattern emerged over the 3.3 to 3.6 micrometre wavelength region, and this pattern was found to be independent of the type of microorganism that was looked at. Thus whether we looked at *E. Coli* or dried yeast cells it did not matter. This invariance came as a great surprise. Our newly discovered invariant spectral signature was a property of the detailed way in which carbon and hydrogen linkages were distributed in biological systems.

The original astronomical spectrum of GC-IRS7 of Dayal and Allen was not of high enough wavelength resolution to verify this prediction unequivocally. If astronomy turned up late with a totally different profile, then our model will have been falsified. From our measurements it turned out that this absorption band was intrinsically weak. This means that a very long pathlength through interstellar dust was needed to get a strong positive signal confirming the presence of bacteria.

The observations that were to mark a crucial turning point in this entire story were carried out by Dayal and D. A. Allen at the AAT in May 1981 *after* our experimental prediction was made. The new observations were of a far superior quality because a new generation of spectrometers were used. Dayal sent us his raw data by fax to compare with our new laboratory spectra which had been obtained just months earlier by Al-Mufti in March and April of the same year. After an hour or so of straightforward calculations we were able to overlay the astronomical spectrum over the detailed predictions of the bacterial model and find a staggering fit. This for us was the best possible confirmation of our model, particularly because the experimental data in the comparison was obtained *before* the final astronomical observations became available. A precise agreement between a set of data points and a predicted curve is normally regarded as a consistency check of the model on which the curve is based. Coming as it did after earlier fits of the same biological model to other sets of data, as we discussed earlier, the closeness of this new fit would be hailed as a triumph of the model. But in our case, since the model of bacterial grains runs counter to a major paradigm in science, the situation was otherwise.

A minor refinement of the original bacterial dust model that yielded the best agreements over a wide range of infrared wavelengths required a mixture of microorganisms including a 10% mass contribution from a class known as "diatoms". The importance of this class of microorganism, possessing siliceous polymers in cell walls, was first brought to my attention by Richard B. Hoover of the Marshall Space Flight Center in Huntsville Alabama. Following a visit to Richard Hoover, during which he and his wife Miriam lavished the most gracious and generous hospitality on us — Priya, me, and our two young children — we published a paper[20] entitled "Diatoms on Earth, Comets, Europa and in Interstellar Space". This represented one of the earliest comprehensive papers in the

---

[20] R.B. Hoover, F. Hoyle, N.C. Wickramasinghe, M.J. Hoover and S. Al-Mufti, 1986. Diatoms on Earth, Comets, Europa and in Interstellar Space, *Earth, Moon and Planets*, **35**, 19–45.

burgeoning science of astrobiology, and it certainly marked the entry of Richard Hoover into this exciting new field. The importance of including diatoms into a cosmic microbial mix was that they represent a class of microorganism that is exceedingly ubiquitous on Earth, and by inference must be expected to be ubiquitous in the Universe.

We were told by a number of chemists with experience of infrared spectroscopy that spectra similar to what we found for our microbial mixtures could be obtained from non-biologically derived organic materials. Since by now we had examined without success literally hundreds of infrared spectra of organic compounds, we did not believe this claim. Consequently, we asked the chemists in question that an explicit example be produced. But it never was, with the exception perhaps that some expensive laboratory experiments, involving carefully controlled irradiation of inorganic mixtures, were claimed to yield undefined "organic residues" that may possess some of the desired properties.

We all to some degree tend to think when we run into an apparently absurd proposal, that any form of opposition to it will suffice. Because in the end what is absurd will be proved to be absurd, so that whatever we say in our opposition will eventually come out on top in the argument. Experience shows that when there are no good observations in favour of what seems absurd, this easily adopted policy is usually fairly safe. But in the face of good observations and in the face of many of them it is a highly questionable strategy.

So just how good is the agreement we had found between theory and data? By the early 1980's, when we attempted to answer this question, we had two decades of experience behind us in evaluating such correspondences. Expressed quite simply, we had never seen anything nearly so good. Yet even so was it all good enough to sustain a belief in such an apparently outlandish idea?

We recognised that to a person who had not followed the problem over the years the observed absorption characteristics of the interstellar grains might have seemed inadequate support for such a far-reaching hypothesis. And doubtless this was the way it appeared to many. But to us who had been involved over almost two decades it seemed otherwise, and we think it fair to add that time has supported our point of view here. Nobody among the critics of the 1980's has managed to find an alternative theory of the absorption characteristics of the grains to equal the success of the bacterial hypothesis. All the spectral features of dust that have recently been observed over a wide range of wavelengths is best explained as the by-products of biology with no other hypothesis needed to paper over any cracks. Prebiology, that is the fashionable explanation these days has no secure basis in fact, and cannot fit the data unless many unjustifiable hypotheses are added.

A new generation of infrared space telescopes (e.g. the Spitzer Space Telescope) have only served to provide more data that supports the cosmic theory of life. From 1980 onwards infrared observations accumulated that also had a bearing on aromatic molecules (molecules involving hexagonal carbon ring structures) in interstellar space. In the mid-1980's groups of astronomers both in the USA and France independently concluded that certain infrared emission bands occurring widely in the galaxy and in extragalactic sources are due to clusters of aromatic molecules. The molecules absorb ultraviolet starlight, get heated for very brief intervals of time, and re-emit radiation over certain infrared lines, including one at 3.28 micrometres. Needless to say, such molecules are part and parcel of biology, and their occurrence in interstellar space is readily understood as arising from the break-up of bacterial cells. Fred Hoyle and I showed much later that galactic infrared emissions at 3.28 micrometres and other well defined infrared wavelengths, combined with extinction at 2175Å can be explained most elegantly on the basis of an ensemble of biologically generated organic molecules.[21]

As was pointed before, even as early as 1962, the presence of aromatic molecules in space might have been inferred from the so-called diffuse interstellar absorption bands. It has been known for over half a century that some 20 or more diffuse absorption bands appear in the spectra of stars, the strongest being centred on the wavelength 4430Å. Despite a sustained effort by scientists over many years no satisfactory inorganic explanation for these bands has emerged. I came across a possible solution at the conference in Troy New York to which I have already referred. F. M. Johnson showed that a molecule related to chlorophyll-magnesium tetrabenzo

---

[21] F. Hoyle and N.C. Wickramasinghe, 1989. *Astrophys. Sp. Sci.*, 154, 143–147; N.C. Wickramasinghe, F. Hoyle and T. Al-Jabory, 1989. *Astrophys. Sp. Sci.*, 158, 135–140.

porphyrin — has all the required spectral properties. Chlorophyll of course is an all important component of terrestrial biology — it is the green colouring substance of plants, the molecule responsible for photosynthesis, the process that lies at the very base of our entire ecosystem on the Earth.

Very recently we have unearthed yet another property of biological pigments such as chlorophylls, a property that clearly shows up in astronomy. Many biological pigments are known to fluoresce, in the fashion of pigments such as exist in glow worms. They can absorb blue and ultraviolet radiation and fluoresce over a characteristic band in the red part of the spectrum. For some years astronomers have been detecting a broad red emission feature of interstellar dust over the waveband 6000–7500 Ångstroms. Chloroplasts containing chlorophyll when they are cooled to temperatures appropriate to interstellar space fluoresce precisely over the same waveband.[22] Cosmic biology announces itself once again in the form of fluorescing pigments, similar to those found in many biological systems.

---

[22] F. Hoyle and N.C. Wickramasinghe, 1996. *Astrophys. Sp. Sci.*, 235, 343–347.

CHAPTER 14

MICROFOSSILS AND EVOLUTION

In 1979 Priya brings out her first book *"Spicy and Delicious"*, published by J. M. Dent, the same publisher who published the books *Lifecloud, Diseases from Space and Evolution from Space*. With this publication she launches into a career as a writer of cookery books and an exponent of Indian and Sri Lankan cuisine. She later goes on to take part in the TV series "Farmhouse Kitchens" and wins a national competition organised by the *Independent* newspaper in London to become Cordon Bleu Cook of the year in 1992. Her cookery writing continues through a series of books leading up to her authorship of *"Leith's Book of Indian and Sri Lankan Cooking"*, and her appointment as Visiting Lecturer and Demonstrator at the prestigious Leith's School of Food and Wine in London. So her aborted career in Law now takes an unlikely turn into the field of gastronomy!

Hoyle and I were now, by the early 1980's, firmly committed to the view that an immensely powerful cosmic biology came to be overlaid on the Earth from the outside some 4 billion years ago, through the agency of comets. Other planetary bodies, within the solar system and elsewhere, must also be exposed to the same process. Wherever the broad range of the cosmic life system contains a form of life (genotype) that matches a local niche of a recipient planet, that form would succeed in establishing itself. In our view the entire spectrum of life on Earth, ranging from the humblest single-celled life forms to the higher animals, must be introduced from the external cosmos. With this in mind we began to examine new data on the planets of our solar system obtained from Pioneer and Voyager spacecraft for tell-tale signs of microbial life. We took as an indispensable condition for bacterial life the need to have access to liquid water.

Subject to this constraint we discovered what we considered were tentative signatures of bacterial life in the planets Venus, Jupiter and Saturn. Since Venus is exceedingly hot at ground level (about 450°C) it would be impossible for life to exist at the surface. Venus, however, has an extensive cloud cover and it is within these clouds that life may have taken root. Water is present in small quantities and in the higher atmosphere the temperature is low enough for water droplets to form. Moreover the clouds of Venus are in convective motion in the upper atmosphere, which ranges in height between 70 km to 45 km, with a corresponding temperature range of 75°C at the top to −25°C at the bottom. We argued that the survival of bacteria over the range of conditions in the upper atmosphere was possible, and that repeated variations of temperature in a circulating cloud system would tend to favour bacteria capable of forming sturdy spores. We argued for an atmospheric circulation of bacteria on Venus between the dry lower clouds and the wetter upper clouds where replication might take place. We discovered that Pioneer spacecraft data, including the presence of a rainbow in the upper clouds, could be interpreted as implying the presence of scattering particles that had precisely the properties appropriate to bacteria and bacterial spores.

These ideas have subsequently come into vogue. In 2002 Dirk Schulze-Maluch and Louis Irwin looked at data on Venus from the Russian Venera space missions and the US Pioneer Venus and Magellan probes. They discovered trademark signs of microbial life from studies of the chemical composition of Venus's atmosphere 30 miles above the surface. They expected to find high levels of carbon monoxide produced by sunlight but instead found hydrogen sulphide and sulphur dioxide, and carbonyl sulphide, a combination of gases normally not found together unless living organisms produce them. They conclude that microbes could be living in clouds 30 miles up in the Venusian atmosphere, exactly in the manner we discussed 25 years earlier (*New Scientist*, 26 September 2002).

We had also argued for localised bacterial populations in Jupiter's atmosphere that might even have a controlling effect on its meteorology, including the persistence of the Great Red Spot. A kilometre-sized cometary object hitting Jupiter at high speed will be disintegrated into hot gas that would form a diffuse patch similar to the Great Red Spot. Such a region of the atmosphere would be rich in the inorganic nutrients needed for the replication of microorganisms. A large bacterial population could then be built up in this area and the possibility arises for a feedback interaction to be set up between the properties of the local bacterial population and the global meteorology of Jupiter as a whole. Additionally we presented a case for bacterial grains trapped in the rings of Jupiter, Saturn and Uranus, rings that were discovered in 1979 by the Voyager missions.

We had also come to regard the presence of methane and other organic compounds in any quantity on solar system bodies as being an indication of life. Essentially all the organics on the Earth today are either directly or indirectly due to biology. So it is likely to be for organic matter that is found in substantial quantity elsewhere in the Universe. The outer planets, which are known to contain methane and other complex organic molecules, in their atmospheres, must be teeming with microorganisms according to our point of view. All these speculations were discussed at length in a Cardiff Blue Preprint entitled "On the Ubiquity of Bacteria" and were later published in abridged form in *Space Travellers: The Bringers of Life* (University College, Cardiff Press, 1981).

The next individual who enters my story is Hans Dieter Pflug from the Geological Institute of Justus Liebig University in Giessen, Germany. In 1979 Pflug had presented evidence for microbial fossils in the sedimentary rocks of South-West Greenland (the Isua Series). These rocks being dated at 3800 million years put the first appearance of life back by some 500 million years from previous estimates, thereby reducing the time available for the development of any primordial soup. In fact it turns out that before 3800 million years the Earth was subject to severe cometary bombardment, so that Pflug's microfossils could represent the first respite from these impacts and the first opportunity for life on Earth to survive. This work has been later supported by studies of several other researchers who have pinpointed the oldest evidence of life on Earth to be at a time between 3.8 and 4 billion years ago, at a time of intense bombardment by comets.

Pflug contacted me in 1980 offering information that was even more interesting than terrestrial microfossils. He claimed to find compelling new evidence for bacterial microfossils in carbonaceous meteorites. The historical background to this work is worth recalling before describing Pflug's new finds.

As the name implies, the carbonaceous meteorites, contain carbon in concentrations upwards of 2 percent by mass. In a fraction of such meteorites the carbon is known to be present in the form of large organic molecules. It is generally believed that at least one class of carbonaceous meteorite is of cometary origin. If one thinks of a comet containing an abundance of frozen microorganisms, repeated perihelion passages close to the sun could lead to the selective boiling off of volatiles, admitting the possibility of sedimentary accumulations of bacteria within a fast shrinking cometary body. We can thus regard carbonaceous chondrites (a type of meteorite) as being relic comets after their volatiles have been stripped.

Microfossils of bacteria in meteorites have been claimed as early as the 1930's, but the very earliest claims were quickly dismissed as being contaminants. The story did not end there, however, and the whole argument was revived in the early 1960's. The actors in the new drama included Harold Urey who was one of the greatest geologists of the century. H. Urey together with G. Claus, B. Nagy and D. L. Europe examined the Orguel carbonaceous meteorite, that fell in France in 1864, microscopically as well as spectroscopically. They claimed to find evidence of organic structures that were similar to fossilised microorganisms, algae in particular. The evidence included electron micrograph pictures, which even showed substructure within these so-called "cells". Some of these structures resembled cell walls, cell nuclei, flagella-like structures, as well as constrictions in some elongated objects that suggested a process of cell division. These investigators, like their colleagues before them, became immediately vulnerable to attack by orthodox scientists.

With a powerful attack being launched by the most influential meteorite experts of the day, the meteorite fossil claims of the 1960's became quickly silenced. One of the more serious criticisms that were made against these claims was that the meteorite structures included some clearly recognisable terrestrial contaminants such as rag-weed pollen. But the vast majority of structures ("organised elements") that were catalogued and described were not contaminants. Intimidated by the ferocious attack that was launched against them, Claus reneged under pressure, and Nagy retreated while continuing to hint in his writings that it *might be so*, rather in the style of Galileo's whispered *"E pur si muove"*.

In 1980 Pflug reopened the whole issue of microbial fossils in carbonaceous meteorites. Pflug used techniques that were distinctly superior to those of Claus and his colleagues and found a profusion of cell-like structures comprised of organic matter in thin sections prepared from a sample of the Murchison meteorite which fell in Australia, about a hundred miles north of Melbourne on 28 September 1969. He showed these images to Fred Hoyle and to me and we were immediately convinced of their biological provenance. Pflug himself was a little nervous to publish these results, fearing for his career and anticipating the kind of reaction that was seen in the 1960's. We convinced him to present his work at the out-of-town meeting of the Royal Astronomical Society, held in 1980 in Cardiff, to which I have already referred.

The method adopted by Pflug was to dissolve-out the bulk of the minerals present in a thin section of the meteorite using hydrofluoric acid, doing so in a way that permits the insoluble carbonaceous residue to settle with its original structures intact. It was then possible to examine the residue in an electron microscope without disturbing the system from outside. The patterns that emerged were stunningly similar to certain types of terrestrial microorganisms. Scores of different morphologies turned up within the residues, many resembling known microbial species. It would seem that contamination was excluded by virtue of the techniques used, so the sceptic has to turn to other explanations as disproof. No convincing non-biological alternative to explain all the features was readily to be found.

We kept in close touch with Pflug throughout the period 1980–1983, and on November 26th, 1981 invited him to deliver a public lecture in Cardiff with the title "Extraterrestrial life: New evidence of microfossils in the Murchison meteorite". The talk was introduced by Fred Hoyle and the meeting chaired by Principal Bill Bevan. The audience was stimulated as well as entertained, and the Earth Scientists left in a state of open-mouthed bewilderment.

We now had an impressive amount of evidence that pointed to the organic composition of cosmic dust and also to an origin of terrestrial life that had to be connected to the wider cosmos. I therefore found it exceedingly puzzling to understand the reluctance on the part of the scientific community to accept the facts. Perhaps there was a perception that very much bigger issues were at stake. If the whole of Darwinian evolution was to come under scrutiny there would be a motive to turn away from even the simplest facts that pointed in such a direction. After all the victory of Darwinism over the narrow Judeo-Christian view of creation as exemplified in the famous Huxley–Wilberforce debate was a hard-won affair and the memory of the blood-letting must still linger in our collective consciousness. It is a victory to be cherished at all cost, and smaller truths may need to be sacrificed in the interests of larger perceived goals.

The next major project I undertook during the period 1980–1981 was an attempt to connect cosmic life, viruses and bacteria causing disease and coming from comets, with the evolution of life on the Earth. If life started on the Earth some 4 billion years ago with a comet bringing the first batch of cosmic microorganisms, how did it evolve and diversify to produce the magnificent range of life forms we see today?

It is believed by Neo-Darwinists that the full spectrum of life is the result of a primitive living system being sequentially copied billions upon billions of times. According to their theory, the accumulation of copying errors, sorted out by the processes of natural selection, the survival of the fittest, could account for both the rich variety of life and the steady upward progression of complexity and sophistication from a bacterium to man. This is perhaps a simple representation of Neo-Darwinism, but it encapsulates its essential features. Is this enough to explain all the available facts of biology? When Fred Hoyle and I began to examine the matter our answer turned out to be an emphatic negative, as we have explained in our two books.[23]

In essence our basic argument was simple. Major evolutionary developments in biology require the generation of new high-grade information, and such information cannot arise from the closed-box evolutionary arguments that are currently in vogue. The same difficulty that exists for the origin of life from its organic building blocks applies also for every set of new genes needed for further evolutionary developments. One of the earliest arguments in this context concerned the origin of the set of enzymes needed for a primitive bacterium.

---

[23] F. Hoyle and N.C. Wickramasinghe *Why Neo-Darwinism Does Not Work* (University College Cardiff Press, 1982); *Evolution from Space* (Dent, 1981).

A typical enzyme is a chain with about 300 links, each link being an amino acid of which there are 20 different types used in biology. Detailed work on a number of particular enzymes has shown that only about ten of the links (or sites) must have an explicit amino acid from the 20 possibilities, while the remaining links can have any amino acid. The smallest known autonomous bacterium present on the Earth, *Mycoplama genitalium*, has a genome comprised of about 400 genes. This means that with a supply of all the amino acids supposedly given, the probability of a random linking of them to produce a functioning enzyme system for *Mycoplama genitalium* can be calculated to be as little as 10^{-1000}. This is not a bet one would advise a friend to take. For comparison, there are about 10^{79} atoms in the whole visible universe, in all the galaxies visible in the largest telescopes.

Such statistics convinced us beyond any shadow of doubt that life must be a truly cosmic phenomenon. The first origin of the magnificent edifice we recognise as life could not have begun in a warm little pond here on Earth nor indeed in any single diminutive location in the cosmos. It must have required the resources of a large part of the universe to originate, and thence its spread is easily achieved. Many attempts to convey this were made by Fred Hoyle and me in our writings by means of comparison with everyday situations. One such comparison was that if a population of 500 people were to throw a pair of unbiased dice, the probability of everybody throwing two sixes is of the same order of difficulty as the origin of the set of enzymes needed for a primitive bacterium. Another comparison is that the origin of life from organic molecules on Earth is of a smaller order of improbability than a tornado blowing through a junk yard assembling a fully working Boeing 707!

The alternative to the random assembly of life as a unique, perhaps unrepeatable event in a finite universe is assembly through the intervention of some form of cosmic intelligence. Such a concept would be rejected outright by many scientists, although there is no purely logical reason for such a rejection. With our present technical knowledge human biochemists and geneticists could now perform what even ten years ago would have been considered impossible feats of genetic manipulation. We could for instance splice bits of genes from one species into another, and even work out the possible outcomes of such splicings.

In 2010 a group of scientists led by Craig Venter in the USA have removed the entire genome of the bacterium *Mycoplasma genitalium* and replaced it with a synthetically assembled genome, producing essentially a modified bacterium which is able to replicate with the new artificial genome. It would not be too great a measure of extrapolation, or too great a license of imagination, to say that a cosmic intelligence that emerged naturally in the Universe may have designed and worked out all the logical consequences of our entire living system.

As we saw earlier our views on cosmic evolution must connect also with the idea of disease-causing viruses coming from space. One might thus legitimately ask: if virus infections are bad for us why did the evolution of higher life not develop a strategy for excluding their ingress into our cells. Logically it seems easy enough for the greater information content of our cells to devise a way of blocking the effects of the much smaller information carried by a virus, and yet this has not happened in the long course of evolution. Could it be, we wondered, whether this "invitation" to viruses was retained for the explicit purpose of future evolution? It is only many years later that an affirmative answer to this question was provided by data from the human genome project. A large fraction of the human genome contains viral sequences that are copied faithfully generation to generation, and this could be the potential for future evolution. Moreover, there is further evidence from genome studies that our ancestral line was attacked periodically with bacterial or viral infections that nearly culled the evolving line save for a small breeding group that came through to modern times.[24]

---

[24]N.C. Wickramasinghe, 2012, DNA sequencing and predictions of the cosmic theory of life, *Astrophysics & Space Science* 2012. DOI: 10.1007/s10509-012-1227-y

CHAPTER 15

ARKANSAS AND THE COLOMBO CONFERENCE

The year 1981 marked the birth of our younger daughter Janaki. Two decades afterwards Janaki obtained a first class honours degree in Mathematics at Bath University — a third generation mathematician in the family — and then went on to do a PhD in astronomy at Cardiff under the supervision of Bill Napier. She enters my scientific scene at a later stage in this narrative, by collaborating on projects that led to an understanding of the processes by which genes of evolved life could be exchanged between planetary habitats throughout the galaxy.[25]

In this year (1981) we had published our book *Evolution from Space* which was receiving a great deal of media attention, particularly a chapter with the enigmatic heading "Convergence to God?" On March 19, 1981, the Governor of Arkansas had signed into law an Act which stated: "Public schools within this State shall give balanced treatment to creation-science and to evolution-science." The US Federal Government challenged the constitutional validity of this Act, and a case was pending between the State of Arkansas and the Federal Government. In view of our much publicised views on the inadequacy of neo-Darwinism to explain the origin of life and evolution, the events that were now to unfold were not entirely unexpected.

In late October of 1981 I received a phone call from Mr. David Williams, the State Attorney for Arkansas to explain the nature of the

---

[25] J.T.Wickramasinghe, N.C. Wickramasinghe, W.M. Napier, *Comets and the Origin of Life* (World Scientific, 2010)

forthcoming trial and inviting me to come as an expert witness for the State. As I understood the situation, a State Education Act No. 590 which required a balanced treatment for "Evolution Science" and "Creation Science" was being challenged by the American Civil Liberties Union as infringing the First Amendment of the Federal U.S. Constitution, an Amendment which required a strict separation of State and Church. Although I held no brief for any particular religion or ecclesiastical group, my sympathies instantly went out to Mr. Williams, both in regard to defending freedom and also because I had come to acquire a dislike for the way that Darwinian evolution was being taught as though it explained *everything* about the nature of life. Darwinian evolution certainly could not explain the origin of life from non-life, at any rate on the Earth, and any opportunity to challenge the established position in this regard seemed inviting.

After speaking at length to the State Attorney I became convinced that all I was required to do in the trial was to defend the ideas we had published in our book *Evolution from Space*. To be their expert witness I had to rebut the claim of the American Civil Liberties Union that neo-Darwinian evolution was in every respect a proven fact. Although I was a little apprehensive, I did not see an immediate reason for declining their invitation. After several long telephone conversations with Fred Hoyle, whose judgement I respected, we agreed that I should go to Arkansas and present a testimony that we would agree upon beforehand. I had religious friends from many faiths and I respected peoples' freedom to hold their particular beliefs, especially if the beliefs were benign. I did not feel that their legitimate aspirations of religious communities of the State of Arkansas should be thwarted on scientific or pseudo-scientific grounds that seemed insecure.

Priya, Janaki and I set out for Arkansas on a cold December day in 1981 through a snowbound airport at Heathrow. There were long delays due to snow and I remember thinking many times that this was an ill omen and that we should turn back and return home. But we made the trip and eventually reached Arkansas just in time for the deposition and the trial. The case I presented essentially summarised my own scientific beliefs. The following quotations are an extract of my testimony.

"The facts as we have them show clearly that life on Earth is derived from what appears to be an all pervasive galaxy-wide living system...Life was derived from, and continues to be driven by, sources outside the Earth, in direct contradiction to the neo-Darwinian theory that everybody is supposed to believe... It is stated according to the theory that the accumulation of copying errors, sorted out by the process of natural selection, the survival of the fittest, could account both for the rich diversity of life and for the steady upward progression from bacterium to Man...We agree that successive copying would accumulate errors, but such errors *on the average* would lead to a steady degradation of information...This conventional wisdom, as it is called, is similar to the proposition that the first page of Genesis copied billions upon billions of time would eventually accumulate enough copying errors and hence enough variety to produce not merely the entire Bible but all the holdings of all the major libraries of the world...The processes of mutation and natural selection can only produce very minor effects in life as a kind of fine tuning of the whole evolutionary process...

In our view every crucial new inheritable property that appears in the course of the evolution of species must have an external cosmic origin... We cannot accept that the genes for producing great works of art or literature or music, or developing skills in higher mathematics emerged from chance mutations of monkey genes ... If the Earth were sealed off from all sources of external genes: bugs could replicate till doomsday, but they would still only be bugs ..."

The notion of a creator placed outside the Universe poses logical difficulties, and is not one to which I can easily subscribe. My own philosophical preference is for an essentially eternal, boundless Universe, wherein a creator of life may somehow emerge in a natural way. My colleague, Fred Hoyle, has also expressed a similar preference. In the present state of our knowledge about life and about the Universe, an emphatic denial of some form of creation as an explanation for the origin of life implies a blindness to fact and an arrogance that cannot be condoned."

My testimony which was consistent with my beliefs, and which Fred wholeheartedly endorsed, is not a source of regret in itself. The State of

Arkansas Education Board that I was representing lost their case. In his summing up of the judgement on 5 January 1982, Judge William R. Overton made the following statement:

> "In efforts to establish "evidence" in support of creation science, the defendants (The State of Arkansas) relied upon the same false premise... i.e., all evidence which criticized evolutionary theory was proof in support of creation science...While the statistical figures may be impressive evidence against the theory of chance chemical combinations as an explanation of origins, it requires a leap of faith to interpret those figures so as to support a complex doctrine which includes a sudden creation from nothing, a worldwide flood, separate ancestry of man and apes, and a young earth...
>
> The defendants' argument would be more persuasive if, in fact, there were only two theories or ideas about the origins of life and the world... Dr. Wickramasinghe testified at length in support of a theory that life on earth was "seeded" by comets which delivered genetic material and perhaps organisms to the earth's surface from interstellar dust far outside the solar system...While Wickramasinghe's theory about the origins of life on earth has not received general acceptance within the scientific community, he has, at least, used scientific methodology to produce a theory of origins which meets the essential characteristics of science.
>
> The Court is at a loss to understand why Dr. Wickramasinghe was called in on behalf of the defendants. Perhaps it was because he was generally critical of the theory of evolution and the scientific community, a tactic consistent with the strategy of the defence. Unfortunately for the defence, Dr. Wickramasinghe demonstrated that the simplistic approach of the two model analysis of the origins of life is false. Furthermore, he corroborated the plaintiffs' witnesses by concluding that "no rational scientist" would believe the earth's geology could be explained by reference to a worldwide flood or that the earth was less than one million years old."

The repercussions of my court appearance unfortunately lasted for several years. Although I had not compromised my beliefs when cross-examined, I often had to agree with the plaintiffs' claims. Many scientists were angry at what they wrongly perceived as our attempt to give credibility to "creation science" which had come to be regarded as the antithesis to science. It was only after meeting "creation scientists" in Arkansas who believed in the literal truth of the Bible, including a belief in an Earth no older than 6000 years, that I began to doubt the wisdom of my decision to testify. For many years following the trial, my family and I were plagued by death threats from several unknown extremist groups, and together with Fred Hoyle I bore a heavy burden of ostracism for expressing our views in the Arkansas trial.

It was a great relief at this time to be able to escape from such troubles. From the beginning of 1980 I became involved in academic and scientific affairs in Sri Lanka in my capacity as advisor to President J. R. Jayewardene. In this connection I had to make frequent brief visits to the island, which were of course quite welcome. President J. R. Jayewardene, as I said earlier, was my father's contemporary at school and had sought me out through the extensive publicity I was receiving at the time for my work on cosmic life. He had visited India and had been much impressed by the standards of scientific research that prevailed there. As newly appointed President he set himself the task of revitalising Sri Lankan science. He invited me in the summer of 1980 to work on a blueprint for an Institute of Fundamental Studies, based roughly on the model of the Tata Institute in India, with the difference that the President himself was to be the Chairman of its Board of Management. By January 1983 the Institute was set up and I was invited by "JR" (as he was called) to be its founding Director, whilst still retaining my position in Cardiff.

I went to Sri Lanka in the summer of that year with Priya and our three children and prepared for our longest stay in the country since our marriage in 1966. We rented an apartment in Colombo and made every effort to reintegrate with the social and academic scene in the island. I had secured a substantial grant from the United Nations Development Programme (UNDP) to put the fledgling Institute on the world scientific map by organising an interdisciplinary international conference under its aegis. The aim of the conference was to introduce prominent scientists from around the world to the Sri Lankan scientific scene in the hope that they would be able to forge links with researchers working in the island, and initiate new research programmes.

Once I had successfully got together the basic infrastructure of the Institute — rented a building with offices located just outside the heart of the city, hired a secretary, an accountant and other office staff — I started work on planning the conference. I proceeded to do this by involving local academics in Sri Lanka. How much more interdisciplinary could one get that the areas of research that I was currently engaged in? I thus found a rationale for including an extended session relating to our immediate research interests. Our list of invitees included Zdnek Kopal, Gustav Arrhenius (grandson of Svante Arrhenius), Hans Pflug, Bart Nagy (who was involved in the controversy over microfossils in meteorites), Keith Bigg (an atmospheric physicist who had found bacteria-like structures in the stratosphere), Sri Lankan expatriate scientists Cyril Ponnamperuma and Asoka Mendis, Phil Solomon, Arthur C. Clarke, Tom Gehrels, Jayant Narlikar and Arnold Wolfendale (then President of the Royal Astronomical Society).

The participants began to arrive a few days before the conference in December 1982 and were all accommodated at the Lanka Oberoi (now the Cinnamon Grand Hotel), a 5 star hotel in Colombo. The meeting itself was to take place at the Bandaranaike Memorial International Conference Hall (BMICH) a well-equipped auditorium and conference facility that had been gifted to Sri Lanka by the People's Republic of China. Fred Hoyle arrived in Colombo from Sidney after an extended visit to Australia. The previous summer we had prepared a preprint entitled "Proofs that Life is Cosmic" which contained what we considered to be the arguments from many different disciplines, all pointing to the cosmic origins of life. We had decided to publish this document as a *Memoir of the Institute of Fundamental Studies, Sri Lanka, No. 1.* The conference itself was a high profile national event in Sri Lanka with President Jayawardene delivering an opening address, followed by a keynote lecture given by Fred Hoyle. Fred gave as usual a brilliant exposition of our theories in the unusual circumstance in which a Head of State, President Jayawardene, was in the audience.

One might have thought that the conjunction of talks by Pflug on the Murchison microfossils, Bigg on microbes in the upper atmosphere, Nagy on D/L ratios of amino acids in meteorites in the presence of Gustav Arrhenius (grandson of Svante Arrhenius), may have struck a chord of consonance. But this was not to be. Hans Pflug cautiously presented his slides as he had done on several occasions in the past, and stated the barest of facts without making any inferences. Likewise Bigg presented his intriguing pictures of particles resembling bacteria in the atmosphere, shyly and with minimal commentary. No sooner than these presentations were completed, Gustav Arrhenius then set upon both them with vengeance, claiming that all such finds had to be interpreted as non-biological artefacts. It seemed that he was determined to turn his back on grandfather Svante's old ideas of panspermia that he considered to be wrong and improper. There was clearly no way of winning him over, no matter how strong the arguments in our favour might turn out to be. People do indeed see what they want to see and are blind to the things they wish not to see.

Bartholomew Nagy's case was different. He knew that he had discovered something profoundly important, first with the microfossils and later with the D/L ratios of amino acids in meteorites. In his formal presentation in Colombo his delivery was strident and straightforward, but when we met him in the hotel bar and engaged him in conversation it was evident that he was an exceedingly frightened man. As Fred Hoyle put it "like a rabbit who was being hunted."

The sessions that dealt with the question of cosmic origins of life were concluded without any resolutions of the main issues that were raised. Arnold Wolfendale (then President of the Royal Astronomical Society) who was a silent observer of these sessions assured us that he would arrange a discussion meeting at the RAS to continue the debate.

Apart from the arguments and conflicts over the life issue, the conference was enjoyed by everyone. The traditionally lavish Sri Lankan hospitality seemed to surpass itself and there was plenty of time for socialising and relaxing. On the final weekend, a fleet of black Mercedes Benz cars provided by the Foreign Ministry, arrived at the hotel to take participants on a three-day cultural tour of the island.

A diversion relating to Cardiff University is in order at this point. It was during this conference that I introduced Bob Churchhouse, then Professor of Computing Mathematics at Cardiff to Ranil Wickramasinghe, the Sri Lankan Minister for Higher Education. The result was that University College, Cardiff benefited from a decade-long link with Colombo by having a string of Sri Lankan students come over to do their PhD's in Cardiff's Computing Mathematics Department.

When all our visitors had left I spent a few more months in Sri Lanka trying to find ways of making the Institute of Fundamental Studies a long-term success. With local jealousies, amongst other factors, this goal was turning out to be more difficult than I had expected. Our time in Sri Lanka effectively ended in July 1983 when the most savage communal riots in the country's history unexpectedly broke out. I was in audience with President Jayawardene at the time when news broke of looting and arson in the South of Colombo, and our meeting was abruptly terminated. The riots evidently began as retaliation for a Tamil Tiger ambush of an army patrol in the north of the country leaving 13 soldiers dead. The Sinhalese retaliated violently and riots continued throughout the island for several weeks.

Tamil Tigers were seeking to establish a separate Tamil homeland in the North, had carried out sporadic acts of violence and suicide bombings mainly directed upon Government targets for over two decades until this time. They were responsible for the murders of several prominent politicians, including Rajiv Gandhi, the Indian Premier, President Premadasa (who succeeded J. R. Jayewardene as President) and several ministers of state in the Sri Lankan Government, including Lalith Athulathmudali to whom I had already referred. The war has now ended with the defeat of the Tigers in May 2009, but in 1983 this was not a country which one could have felt comfortable to work in.

CHAPTER 16

THE LECTURE AT SLBC

With my elevated status in Sri Lanka as an advisor to the country's President and the Director of the Institute of Fundamental Studies, the years 1980–1981 were filled with a variety of social and semi-social engagements. One such engagement was as chief guest at the prize giving at St. Thomas College, a boys school that was the arch rival to my old school Royal College. These two schools, like Eton and Harrow vied with each other for supremacy in sport as well as in scholastic achievements. The annual Royal-Thomian Cricket match was the highlight in the country's sporting calendar. So speaking at my rival school's Prize day was a rather strange experience.

Another event of note was an invitation I received from the SLBC (Sri Lanka Broadcasting Corporation) to give an inaugural lecture in a new series "Expanding mind". I accepted this gladly because I have always felt that dissemination of discoveries in science to the general public (outreach as it is now called) is an important role of the scientist. This lecture is online at http://www.youtube.com/watch?v=gGVaOdA9IwE but because it is part of my scientific story and describes the state of play of various parts of it as of 1980, I shall reproduce extracts from it below, even at the risk of reiterating things I have covered in earlier chapters.

"…..In this lecture I shall discuss the question of the origin of life on our planet in a somewhat wider context than is usually thought to be necessary. First of all, we must concede that there is a logical need to understand an origin of life, at least as far as the Earth is concerned. The Earth, today, is surely teeming with life, and organic molecules exist in great abundance. But this, clearly, was not always so. Four-and-a half billion years ago the material of the Earth was in the form of a cloud of fine dust particles that was in the process of contracting to become the primitive Earth. And for a while after the Earth had itself condensed into a solid body, its molten crust would have been too hot for any organic molecules to persist.

There are good reasons to believe that much smaller condensed bodies in the form of icy comets from the outer regions of the solar system subsequently impacted on this cooling planet and deposited volatile materials, including water, that went to form the Earth's oceans. Evaporation of water from the oceans, and the break-up of water molecules by sunlight then gave rise to an atmosphere and a cloud cover around our planet. Only after this happened could the Earth have become a suitable home for life, with its surface screened and well protected from the damaging ultraviolet radiation from the Sun.

We shall next touch briefly on a question central to the theme of this lecture: could life have emerged from non-living matter here on the Earth? The production of organic molecules of rather great complexity — for example, sugars, nucleotides, amino acids — is a necessary condition for the origin of life; but the production of these substances alone are by no means enough for the origin of life. The vast majority of the organic molecules we find today on the Earth are either directly or indirectly the product of biology.

There are of course abiotic, or non-biological conditions, that could lead to the production of organic molecules from inorganic ones, and there have been several classic experiments to show this. Perhaps the most famous of all is the Urey–Miller experiment which was carried out nearly thirty years ago. In this experiment a flask filled with H_2, H_2O, CH_4, in appropriate proportions was sparked with an electric discharge, and it was found that traces of biochemical substances were formed. This result, showing as it did that the basic chemical building blocks of life could be produced non-biologically, appeared at the time so dramatic that it had a profound psychological impact on the world of science. Many biologists were led to believe quite firmly that they had come near to understanding an origin of life in terms of processes that might have taken place here on the primitive Earth. Yet a few odd amino acids and sugars produced under laboratory conditions comes nowhere near the exceedingly intricate complexity of life itself. And it is a far cry from proving that life could ever have started here on the Earth.

On the contrary, a terrestrial origin of life would, for many reasons, seem most unlikely. The conditions on the primitive Earth are not likely to have been appropriate even for the production of the building blocks of life — let alone for the origin of life. And even if the chemical building blocks of life were supplied — God given, say — in some terrestrial pond, their assembly into life would be well-nigh impossible. The Earth is too small, the available time-scales too short, and life is far too complex for this to have happened.

Many biologists who think that an origin of life took place here on Earth implicitly favour a miracle. The argument goes as follows. Given all the right starting materials, the origin of life is such an improbable process, occurring with such an extremely low probability, that it could, perhaps, have happened only once in the entire Universe. And — the argument goes on — since we know life is present here, it must have started here. This second step does not follow logically, particularly if we admit the possibility of transferring micro-organisms such as bacteria from one part of the galaxy to another. The Earth-centred view of life that most scientists hold is essentially pre-Copernican, and its adherence leads to many serious difficulties in understanding and interpreting biological data.

Let us now turn to evidence. The first sedimentary rocks on the Earth were laid down through the effects of rainfall and water erosion of an initial crust at about 3.8 billion years ago. Until quite recently it was thought there was about half a billion years between the laying down of the first sedimentary rocks on Earth and the start of life. And the origin of life was thought up to quite recently to be recorded in the form of microfossils of bacteria and algae in the Swaziland cherts — rocks in a certain mountain formation in South Africa dated at about 3.3 billion years ago. If this were true there might have been about a half billion year time span for some kind of primordial soup to have brewed on the Earth, at any rate according to the conventional theory.

This seemingly comfortable situation for the conventional theory has now been dramatically shattered. It has recently been discovered that there are the most unambiguous signs of life in rocks from the Isua region of West Greenland that have been dated at 3.83 billion years ago, rocks that represent, perhaps, the very first record of sedimentary processes on our planet. This result has been confirmed quite independently by two groups of scientists, one American, another European.

Cyril Ponnamperuma and his colleagues at the University of Maryland have found unambiguous evidence for the chemical residues of photosynthetic micro-organisms in organic material extracted from these rocks. H.D. Pflug and H. Jaeschke-Boyer who also studied rocks of the same age found even more dramatic results (Nature, vol 280, p. 483, 1979). Not only were there the chemical residues of living material, as found by Ponnamperuma and his colleagues, but these were found to be localised quite remarkably, within organised structures that closely resembled present-day yeasts. The analysis was done using very thin slices of rock in which the fossil cells were preserved and held in a way that ruled out any chance of more recent biological contamination.

The primordial soup seems to have just about been squeezed out of existence from the geological record. There is now evidence that the Earth was showered with living cells from the very dawn of its creation. Life got a toehold on the Earth at the first moment that physical conditions became favourable. A hitherto sterile Earth might be said to have become infected with life, a life which thereafter began to evolve against the background of continually changing local conditions on the Earth. And with initial-life forms that were as complex as yeast cells, terrestrial evolution need have involved little more in essence than an unravelling of a wide range of cosmically determined possibilities.

If life did not start here we might ask the question: where did it first start? The number of possible sites for where life could have first started are of course legion. In our own solar system we can argue that the conditions in the interiors of any one of a 100 billion comets are better suited to this event than conditions that could ever have existed on the Earth itself.

And with more than a hundred million sun-like stars in the galaxy, the probability of life first starting on any comet in the solar system would be so miniscule as to be totally negligible.

Life could have first started with equal probability on any one of 10^{22} comets in the galaxy. So to assert that this event occurred on an object within our own solar system must be regarded as almost arrogantly ego-centric. Bacteria have sizes that make it difficult for them to remain confined in orbit around a star. Starlight has the property of being able to expel particles of bacterial sides away from the gravitational attraction of a star. Bacterial cells are thus explosively propagated throughout the galaxy, no matter where their first origin occurred.

Bacteria on the Earth have properties that are difficult to reconcile with the usual concept of a terrestrial origin and evolution of bacteria. The total mass of bacteria on the Earth, mainly in the soil and the sea bed, is about ten billion tons. The number of bacterial species involved here is probably vast. One of the most striking features about the distribution of terrestrial bacteria is that they are rarely optimally matched to the environments in which they are found. If bacteria are indeed indigenous to the Earth one would expect an almost precise adaptation of organisms to niches. And this is not found. For example, the distribution of two classes of bacteria — the heat-loving bacteria and the cold-loving bacteria — presents a continuing puzzle. A fraction of bacteria in tropical soils are of the cold-loving type, bacteria that have specialised abilities to multiply at temperatures below freezing-temperatures that are never realised in the tropics. Conversely, a fraction of the bacteria in the Arctic and Antarctic ice shelves are of the heat-loving sort. These properties are readily understood if the Earth is bombarded with an astronomically vast range of different bacterial types. The various niches on the Earth pick up the types that can survive under the conditions that locally prevail, but which are not necessarily well-adapted to those conditions.

There are other properties of bacteria that can be regarded as being distinctly unearthly. For instance, bacteria have the property of being able to survive almost indefinite periods of time under the low temperature conditions of interstellar space. Bacteria can withstand and survive doses of ultraviolet radiation that are never received on the Earth. Some bacterial types have an uncannily high resistance to X-rays, gamma-rays and cosmic rays.

The most crucial astronomical clues have come from a study of interstellar clouds — clouds of obscuring material that are known to exist in the space between stars. Interstellar clouds show up as conspicuous dark patches and striations against the background of stars along the Milky Way. Several generations of astronomers have pondered on the composition of material within such clouds. A large fraction of the mass of the entire galaxy resides in the form of gaseous molecular hydrogen in these clouds. It has recently been found that large quantities of organic molecules are also present — in all about three dozen or so gaseous organic molecules have been discovered to date. But perhaps the most baffling component by far is a population of tiny microscopic dust particles that inhabit the clouds, particles that cause scattering and obscuration of the

light from distant stars. For nearly 20 years we have strived to find a particle which has the astronomically observed properties of these so-called
grains in space. But we had little or no success. Then, about six months
ago, we had the rather outlandish idea to try a comparison of these grains
in space with bacteria. To our great surprise we found that terrestrial bacteria are uncannily similar to cosmic dust with regard to all their known
properties. The measure of similarity and agreement was indeed so perfect
that we were led to conclude that the hitherto unidentified component of
interstellar dust clouds were in fact bacterial cells. The total mass of bacterial cells, along with their associated viruses, throughout the galaxy turns
out to be truly enormous, measuring some ten million times the mass of
the Sun. (The long struggle that eventually led me to this conclusion has
been recorded in earlier chapters.)

The picture which has emerged then is that space is filled with living
cells, cells which are mainly in a frozen dormant state. Every niche suitable for life that develops naturally through collapse of cosmic gas clouds
and the formation of stars, comets and planets becomes very quickly
infected with this all-pervasive living system. Such 'infections' lead to a
vast increase in the numbers of bacteria and their inevitable feed-back into
the space between stars. It now becomes almost meaningless to pose the
question: In which star system did life begin? Life from this point of view
is the sum total of an evolutionary experience gained and accumulated in
a multitude of different places, quite widely separated, and extending in
time throughout the entire history of the Universe.

Coming nearer home, in our own solar system, we argue that life in
the form of bacterial cells, viruses and virions, were first housed in the
comets. Every one of nearly a thousand billion comets had warm watery
interiors and all the organic nutrients necessary to make for a congenial
breeding ground for microbial life. We mentioned earlier that comets
crashing on to the Earth brought the oceans and the atmosphere. With the
oceans and the atmosphere so deposited, comets would also have seeded
our planet with life — a life which under the protected canopy of cloud-
covered skies was able to take root and to flourish.

The first successful seeding of the Earth with life occurred at about
3.8 billion years ago. But in our view this process of seeding could not
have stopped at this distant prehistoric time. Comets are with us in the

solar system today, and the Earth continually picks up debris from comets. About 100 metric tons of cometary debris enters the Earth's atmosphere every day. Much of this debris is either sterile due to being around for too long near the Sun, or is burnt up on entering the Earth's atmosphere. But we could argue, that a small fraction of the incoming dust freshly evaporated from comets, must contain active or viable microbes that actually survive entry through the Earth's atmosphere.

This conclusion, bold though it may be, has the advantage of being susceptible to proof or disproof, especially if the Earth is being showered with micro-organisms that are pathogenic to plants and animals. Rather as physicists use amplifying electronic counters to detect very small fluxes of incoming cosmic ray particles, so in a similar way plants and animals could be regarded as amplifying detectors for microbes from space.

A reading of medical history gives ample encouragement for this point of view. Many bacterial and viral diseases have a record of abrupt entrances, exits and re-entrances on to our planet — exactly as though the Earth was being seeded at periodic intervals. In the case of smallpox the time interval between successive entrances appears to have been about 700–800 years. Since man is the only host of the smallpox virus, global remissions of this disease lasting for many hundreds of years are very hard — almost impossible — to understand. From the conventional point of view one has to say that the virus became extinct, and then re-evolved to precisely its original form from some unknown ancestor after many hundreds of years. A most improbable event that would be. But historical data is often incomplete, and one may have doubts about the genuine absence of any disease from the Earth at a given time. We know that smallpox was present a few hundred years before the classical period in Greece, and again, most decisively, that it was present a few hundred years after the dawn of the Christian era. With the accurate medical records that are available at the time we can be equally sure that smallpox was not present in classical Greece and Rome. A genuine world-wide absence would seem to be implied at this time, for otherwise it would be hard to imagine a disease so infectious as smallpox being kept out for so long from the hub of empire in the Western World.

There are many puzzles too in the medical annals of more recent times-puzzles that are resolved if we accept that pathogens could be falling from the skies. For instance, there is a group of about 500 Trio Amerindians who until quite recently had lived in the dense forests of Surinam quite out of contact with the rest of humanity. When the forests were cleared and this tribe was discovered by anthropologists, it was found that there were several polio victims who seem to have contracted this disease at times roughly coincident with epidemics in cities hundreds of miles away. There is no conceivable way by which the forest dwelling Surinam Indians could have contracted polio from city dwellers. But the city dwellers and forest dwellers could both have caught the disease if the disease causing pathogens rained on them from above.

The attack of pathogens from space are of course not confined to man. There is a considerable body of evidence relating to attacks on non-human species as well. For example, a new viral disease began to take a heavy toll on dogs in the year 1978. The most remarkable feature here was that this lethal disease appeared almost simultaneously in widely separated parts of the world. Veterinary surgeons in 22 states of America, in Canada, South Africa, Holland and Australia all noticed the disease to appear for the first time at about the beginning of May of that year. Since there is clearly no traffic of dogs between these countries, and because the most stringent quarantine measures are everywhere quite ruthlessly enforced, this simultaneity of attack is strongly suggestive of an invasion from space.

Perhaps the most impressive evidence of all for diseases from space comes from a study of the incidence of influenza. It has been known for many years that epidemics of flu strike fairly large tracts of a country almost simultaneously, and appear to spread far too rapidly for it to be caused by one victim infecting another. For instance, Dr. Robert Thomas observing the pattern of several epidemics and writing as far back as 1813 had this to say:

"By some physicians influenza is supposed to be contagious; by others not so; indeed its wide and rapid spread made many suspect some more generally prevailing cause in the atmosphere, as alone capable of accounting for its extensive and speedy diffusion"

The records of later epidemics reveal the same general pattern. ....
According to our point of view, reservoirs of the causative agent for influenza are periodically resupplied at the very top of the Earth's atmosphere.

Small particles tend to remain suspended high up in this region for long periods unless they are pulled down into the lower atmosphere. In high latitude countries, such breakthrough processes, where the upper and lower air become mixed, are seasonal and occur during the winter months. Thus a typical influenza season in a European country would occur between December and March. Frontal conditions with high wind, snow and rain effectively pall down viral pathogens close to ground level. The complex turbulence patterns of the lower air ultimately control the details of the attack at ground level, and determine why people at one place and at one time succumb, and why those in other places and at other times do not.

We could now ask the question: is the incoming space pathogen which causes influenza epidemics the fully-fledged virus, or a simpler structure which could in some way serve as a kit to construct the virus? When we first began thinking about the possible extraterrestrial incidence of diseases we had assumed, as was indeed customary in most circles, that a viral disease must necessarily be caused by a fully-fledged virus. This need not necessarily be so. There are well-documented instances of diseases in which viruses are known to be released from within the host cell.

An empirical fact beyond question is that viral particles are exuded from patients suffering most of the well recognised viral diseases. For certain viral diseases such as smallpox, the virus so exuded is highly robust, survives desiccation and exposure to light, and could succeed in infecting other susceptible individuals. In other diseases this is not necessarily so. In the case of influenza an exuded virus does not survive outside a human host for more than a couple of hours under normal conditions. This property was well confirmed in our study of case data from English and Welsh schools, which indicated the lack of any significant degree of lateral transmission.

There is however a dilemma implied here in regard to the requirement that our presumptive space-borne virus must have different properties from those of the exuded virus so as to enable its survival during transit from comet to Earth and the subsequent passage through the Earth's atmosphere. One way to resolve this is to propose that viruses which are primary agents in the infective process come suitably clothed in a protective matrix. For instance, a matrix comprised of organic material such as cellulose, in the form of a particle of diameter — 10^{-5} cm (a tenth of a micrometre), would seem to be ideally suited to the preservation of a virus.

We recently realised that there might be another, perhaps more elegant way to explain the observed facts relating to the incidence of influenza. Small strands of RNA known as viroids are now recognised as being the causative agents in certain plant diseases. It is also known that such viroids could play a role in regulating the translation and expression of the cellular DNA. The logical possibility thus arises where viroids could serve to unleash latent viruses. Viroids have an advantage over viruses as the primary agent that arrives to the Earth from outside, since they are much better equipped to survive and remain viable for long periods under the hostile conditions that exist in space.

Quite apart from the negative role in causing disease, space-borne viroids and viruses could have a positive evolutionary role. We can argue that viral and viroid invasions from outside provided the main evolutionary driving force in biology, the main source of additional information needed for the evolution of life. Viruses add on to our genomes over geological time and virions sporadically unlock hidden genetic potential. Without a steady supply of cometary viruses life on Earth may not have evolved beyond the state of a simple microbe. In its ultimate analysis disease caused by viroids, viruses and bacteria although decisively bad for the individual, is good for a species, and appears moreover to be absolutely necessary for its evolution.

The ideas I have described in this lecture, both singly and taken together, represent radical departures from currently accepted beliefs in science. Scientists as a rule are reluctant to accept such change, but, as we have indicated, these changes of dogma are dictated, almost inevitably, by the facts that have emerged.

Throughout our long history as a thinking species we have always been loathe to accept a theory of the world in which we ourselves did not occupy a central and most important place. The view that the Earth was the centre of the solar system was held for centuries, and was abandoned only with great anguish after Copernicus. We have now come to accept the position that the solar system is not the centre of the galaxy, and that our galaxy itself is not the centre of the Universe.

Likewise, this same Copernican-style revolution must be applied to life. Life, with its extraordinary subtlety of design and its exceedingly intricate complexity, could only have evolved on a scale that transcends the size of our planet, the size of the solar system, even perhaps the size of the entire galaxy.

CHAPTER 17

ARCHAEOPTERYX AND THE RETURN
OF HALLEY'S COMET

By 1983 Cardiff Astronomy was advancing rapidly in many different directions. There was a noteworthy development under Bernard Schutz on Relativistic Astrophysics that led eventually to a major group devoted to the search for gravitational waves — a prediction of Einstein's theory of relativity. In my particular areas of research, which were by now getting increasingly distant from the rest, there were several students, mostly from Iraq, working alongside me and Shirwan Al-Mufti on various aspects of the biological dust grain thesis.

After the heat of the Colombo conference, the next testing ground for our ideas was a discussion meeting of the Royal Astronomical Society (RAS) which took place on 11 November 1983. As promised to us in Colombo, the meeting was initiated by Sir Arnold Wolfendale. The discussion had the title: "Are interstellar grains bacteria?" Apart from myself and Fred Hoyle, the participants included Hans Pflug, Phil Solomon and Mayo Greenberg. After Fred Hoyle and I had presented our evidence in support of an affirmative answer to the question at issue, the others set out to argue the case against. In our view, the case against and the rebuttals were largely polemical.

Almost concurrently with the RAS conference evidence was being gathered for organic molecules and organic structures in extraterrestrial material falling onto the Earth. Don Brownlee[26] had begun his programme of collecting cometary dust particles using high-flying U2 aircraft that swept through the lower stratosphere at a height of 15 km. The method

---

[26] Bradley, J.P., Brownlee, D.E., and Fraundorf, P.: 1984, *Science*, 223, 56.

employed was a "flypaper technique" where a sticky plate swept through large volumes of air and captured aerosols as they struck the surface at high speed. Fragile structures like clumps of bacteria or volatile dust would have unfortunately been destroyed by this procedure. What were recovered were mostly porous siliceous clumps with some embedded organics, but occasionally entire organic structures of cometary origin were found buried within them. When we looked at the published pictures of these structures we found at least one clear case of an embedded bacterium-like organic object with minute embedded magnetite substructures. This structure also turned out to be uncannily similar to a well-recognised fossilised iron-oxidising bacterium in the Earth's sediments dated at 2000 million years. Since the latter was found by Hans Pflug, we got together with him and published this comparison in a paper entitled: "An object within a particle of extraterrestrial origin compared with an object of presumed terrestrial origin".[27] This, in our opinion, was the first strong indication that cometary particles with a biological provenance are still entering the Earth's atmosphere.

In my journey of life I sometimes took diversions that led through treacherous paths. With hindsight some of these adventures may better have been avoided. One such adventure, that extended over three years, was concerned with the famous fossil of *Archaeopteryx*, which is considered to be the link between reptiles and birds — half bird, half reptile. When it was discovered it was hailed as one of the long-sought after missing links in the fossil record. The fossil, ostensibly of a small reptile with exquisitely well-preserved feather impressions, was discovered in a 160 million year old limestone deposit from the Jurassic period. The discovery was made in 1877 by a German doctor Ernst Habelein in a quarry at Solnhofen, some 40 miles south of Nuremberg. A few years earlier the same doctor had provided the palaeontological community with a single feather impression on limestone also of the same age, and from the same quarry. The limestone slab containing the full *Archaeopteryx* fossil, as well as its counter slab (containing the mirror impression of the fossil) were sold to the British Museum, who have prized it as one of their most precious possessions. Little wonder then that

---

[27] F. Hoyle, N.C. Wickramasinghe and H.D. Pflug, 1985. An object within a particle of extraterrestrial origin compared with an object of presumed terrestrial origin, *Astrophys. Sp.Sci*, 113, 209–210.

even the slightest attempt to challenge its authenticity would provoke a furore.

In September 1984 Fred Hoyle received a letter from a certain Lee M. Spetner in Rehovot, Israel, which read thus:

"For several years I have had a strong suspicion that the *Archaeopteryx* fossil is not genuine...I suspect that the fossils were fabricated by starting with a genuine fossil of a flying reptile and altering it to make it appear as if it originally had feathers..."

Both Fred Hoyle and I were both naturally intrigued by the suggestion. We promptly headed to our nearest libraries (the internet was not available yet) to learn all we could about the history of this fossil. It appeared that fossil forgery was quite common in those days, so for museums to be sold forgeries was by no means an absolute impossibility. There were at least three *Archaeopteryx* related fossils on record from the same source: the single feather (just referred to), the specimen at the British Museum, and another similar specimen in Germany. If the one fossil turned out to be a forgery, the likelihood was that they all were.

A few weeks later Spetner sent us a detailed manuscript in which he summarised his concerns about the authenticity of *Archaeopteryx*, and we felt at the time that he had a *prima facie* case for his thesis. I met Spetner shortly afterwards when he visited Cardiff, and I formed the opinion that he was an honest man — an orthodox Jew who lived by the Book. In fact he turned out to be so orthodox that when Priya entertained him to dinner at our house she discovered that he would not eat any food that was cooked in a non-Jewish home. We had to provide him with raw carrots and apples, which was evidently all he was permitted to eat. However, Priya through her network of friends was able to find an orthodox Jewish family who were able to give him the Kosher food that was required.

Without indicating what our motives were, we now approached Dr. A. J. Charig at the British Museum for permission to photograph the fossil. Permission was duly granted. In the afternoon of 18 December 1984 we went over to London with the Physics Department's photographer R. S. Watkins and took hundreds of pictures of both the slab and the counterslab under various lighting and exposure conditions. When we studied the pictures we saw many features that convinced us at this time that Spetner's case was one that had to be taken seriously. The perfect feather impressions looked as though they were impressed on a thinly applied over-layer (limestone cement) that looked markedly different in texture from the rest of the fossil. We also found that the slab and the counterslab did not match in certain crucial areas. Playing the role of amateur detectives we poured over our pictures for hours on end, and after several months we were sufficiently convinced to go into print. We published our pictures with accompanying comments, raising some doubts about the authenticity of the fossil, in several articles written for the *British Journal of Photography*. The Editor of the *British Journal of Photography*, Mr. Crawley, issued press releases highlighting the articles, and as a consequence the whole affair received far more media publicity than we would have wished for.

Our studies of the pictures as well as our delving into the history of this fossil occupied several months and led us to publish perhaps our most controversial book.[28] The Natural History Museum in London is a hallowed national institution and did not like this book and the implied threat to its integrity. The Museum went to the trouble of mounting a public exhibition to set out their case against the forgery claim. Finally they thought a line could be drawn under an ugly interlude in their history when they published a rebuttal in an article in the journal *Science*.[29]

But all this was not enough to convince us. Further tests done some time later by Spetner in Israel on a small sample that he secured from the Museum still left the matter wide open and essentially unresolved. The upshot of all this was that nothing was decisively resolved. We did not convince our opponents as we had set out to do, and we also lost many friends! The prudence of taking on so powerful an institution like the British Museum must in retrospect be called to question. Particularly since the outcome of our intended inquiry, whichever way it went, would have had no bearing whatsoever on the bigger issues at hand.

Fred Hoyle's visits to Cardiff were always major family events. During his stays with us he would find time to discuss matters that were far removed

---

[28] F. Hoyle and N.C. Wickramasinghe, *Archaeopteryx: The primordial bird — A case of fossil forgery* (Christopher Davies Publishers, Swansea, 1986).

[29] A.J.F. Charig *et al.*, 1986. *Science*, 232, 622.

from science and as my children grew older they too came to appreciate his rich and diverse company. Fred had an abiding interest in classical music and on some mornings (he was an early riser), I would find him in the living room listening intently to Beethoven's Fifth Symphony or Mozart's Death Requiem — music that I recall hearing on so many occasions blaring out from a gramophone at 1 Clarkson Close. He had grown up in the midst of music, his mother being a gifted piano teacher who had studied at the Royal College of Music, and he himself a paid chorister at a local church. Our home in Cardiff too tended to be filled with music as Priya and our three children share a passion for music. Whenever my elder daughter, Kamala, played the piano, Fred would stop whatever he was doing and praise her talents. In fact he would often request her to play whilst we were discussing our work.

An amusing incident took place when his birthday (June 24) happened to fall during one of his visits. Priya and I had organised a dinner party to celebrate the event at a restaurant known as "The Walnut Tree", near Abergavenny. Situated at the foot of the Skirrid Mountain and frequented by the famous food writer Elizabeth David, it was arguably the best restaurant in Wales. We had invited a few other astronomers and friends to join us and the occasion turned out to be most memorable. At the end of the evening the waiter brought me the bill to pay. I produced my visa card as I always do on such occasions, and to my horror was informed that they accept only cash or cheques. This moment of embarrassment was aggravated by a request to write down my name and address to be shown to the proprietor for appropriate action. Within minutes an ecstatic Franco Taruschino (owner and chef) rushes to our table and hugs Priya! Evidently he knew Priya from a recent cookery book she had published. Fred Hoyle and all the other astronomers at our table were unknown entities as far as he was concerned. His excitement at seeing Priya was so great that he went back in and woke up his young daughter to introduce her. All's well that ends well — after a round of complimentary liquors, we were sent home as friends of VIP Priya, and asked to post a cheque whenever we had the time!

The next high point in my scientific career was connected with the return to perihelion of Halley's comet in 1986. This was the first time that a comet was being studied by scientists since the beginning of the space age. From as early as 1982 a programme of international cooperation to investigate this comet came into full swing, the aim being to coordinate ground-based observations, satellite-based studies, and space-probe analysis on a worldwide basis. No less than five spacecrafts dedicated to the study of Comet Halley were launched during 1985, the rendezvous dates being all clustered around early March 1986, about one month after the comet's closest approach to the sun.

In the immediate run-up to these events Fred Hoyle and I met to discuss what observations might be likely according to our present point of view. What predictions might we possibly make? Our deliberations led us to conclude that organic/biologic comets of the kind we envisage would have exceedingly black surfaces. This is due to the development of a highly porous crust of polymerised organic particles that can permit vigorous outgassing only when the crust comes to be ruptured. We put all our arguments in the form of a preprint entitled "Some Predictions on the Nature of Comet Halley" dated 1 March 1986 (Cardiff Series, No. 121) which came to be published much later in *Earth, Moon and Planets*, (Vol. 36, 289–293, 1986). This was only twelve days before the encounter, and our priority would have gone unrecorded had it not been for the fortunate circumstance that the *London Times* picked up on it and reported its contents (*The Times*, March 12, 1986).

On the night of March 13, 1986 we watched our television screens with nervous anticipation as Giotto's cameras began to approach within 500 km of the comet's nucleus. The fears that the spacecraft might be badly damaged and even destroyed by impacts with cometary dust were proved to be wrong, and the equipment functioned well throughout the encounter. The cameras were expecting to photograph a bright snowfield scene on the nucleus consistent with the then fashionable Whipple dirty snowball model of comets. In the event the television pictures transmitted world-wide on 13 March proved to be a disappointment. The cameras had their apertures shut down to a minimum and trained to find the brightest spot in the field. As a consequence, very little of any interest was immediately captured on camera — the scene was far too dark. The much publicised Giotto images of the nucleus of Comet Halley were obtained only after a great deal of image processing. The stark conclusion to be drawn from the Giotto imaging was the revelation of a cometary nucleus that was amazingly black. It was described at the time as being "blacker than the blackest coal ... the lowest albedo of any surface in the solar system...." Naturally we jumped for joy! As far as we were aware at the time we were the only scientists who made a prediction of this kind, a prediction that was a natural consequence of our organic/biologic model of comets. Fred and I regarded this development as yet another decisive triumph of our point of view. More triumphs were soon to follow.

A few days after the Giotto rendezvous, infrared observations of the comet were made by Dayal Wickramasinghe and David Allen using the 40 metre Anglo-Australian Telescope (*IUA Circular* No. 4205, 1986). On March 31, 1986 they discovered a strong emission from heated organic dust over the 2 to 4 micrometre waveband. As noted earlier basic structures of organic molecules involving CH linkages absorb and emit radiation over the 3.3–3.5 micrometre infrared waveband, and for any assembly of complex organic molecules as in a bacterium, this absorption is broad and takes on a highly distinctive profile. The Comet Halley observations by Dayal and David Allen were found to be identical to the expected behaviour of desiccated bacteria heated to 320 K. Another victory for our model! Later analysis of data obtained from mass spectrometers aboard Giotto also showed a composition of the broken-up fragments of dust as they struck the detector to be similar to bacterial degradation products.

The Halley observations, in our view, clearly disproved the fashionable Whipple's "dirty snowball" theory of comets. The theory dies-hard, however, with variants of it still in vogue with the claim that Whipple was still mostly right, except that there was more dirt (organic dirt) than snow! It could not be denied that water existed in comets in the form of ice, but great quantities of organic particles indistinguishable from bacteria are embedded within the ice. This conclusion was unavoidable unless one chose to ignore the new facts.[30]

---

[30] D.T. Wickramasinghe, F. Hoyle, N.C. Wickramasinghe and S. Al-Mufti, 1986. A model of the 2–4 µm spectrum of comet Halley, *Earth, Moon and Planets*, **36**, 295–299.

CHAPTER 18

ESCAPE FROM BREWING TURMOIL

There are moments in life one treasures above all else because of the knowledge that they would never return. In my own life's journey one such moment came in the summer of 1987, on the occasion of my parents' golden wedding anniversary. They were visiting the UK during the summer as they regularly did, and my brother Sunitha (then Professor of Heamatology at Imperial College) had arranged a celebration lunch at a hotel near Maidenhead. There were no speeches nor formalities, but together with my three brothers and their families, I savoured this rare moment of personal history with nostalgia and quiet reflection. Our happiest of childhood moments at 35 Hildon Place leapt out through the mists of time, and events in our two years at 23 Pattison Road, Hampstead, in 1947, flashed back with a vividness that seemed almost unreal. As we raised our glasses to our parents I wondered how many similar moments of our own family life, with Priya and our kids, would thus be remembered in years to come. We also reflected on the sobering thought that the many contributions to knowledge in various fields that had been made by the four brothers — Sunitha (medical science), Dayal (astronomy) and Kumar (nanophysics) — were in a sense a fulfilment of our father's own remarkable genius that was never given the opportunity to reach fruition.

In 1986 President J. R. Jayewardene had stepped down from being executive President of Sri Lanka and major changes were afoot in the island. So also was Cardiff in a state of turmoil in the academic year 1987/1988. University College Cardiff, to which I was appointed a tenured Professor and Head of Department in 1973, was deemed to be in financial difficulty and Principal Bill Bevan was forced to resign. Furthermore, University College Cardiff and the neighbouring University of Wales Institute of Science and

Technology were forced to form a single merged University institution within the University of Wales. This new institution came into existence in 1988 under the name University of Wales College of Cardiff and was later called Cardiff University. In the process of merger individual departments in the two constituent colleges were forced to unite. There were four mathematics departments at University College Cardiff and one at the University of Wales Institute of Science and Technology that were all combined into a single School of Mathematics. The large contingent of astronomers in my own Department of Applied Mathematics and Astronomy were now "displaced persons" and were forced to join either a new School of Mathematics or a Department of Physics, and staff had to be distributed between the two new departments. I had a choice between Physics and Mathematics and after consulting Fred Hoyle decided to stay in the School of Mathematics.

This was not a happy time in my professional life at Cardiff University and I have deliberated greatly as to whether I should include in this book the successive discrimination towards me following the merger. Erring towards its non-inclusion, I was reminded of the words of JR Jayawardena in the UN talks following the defeat of Japan when he advised on a non-retaliatory course of action "Hatred shall not be ceased by hatred". However, as this book comprises the story of my professional life, I decided that it should be touched upon briefly, although in reality if I were to document the extent of what happened that would be a book in itself.

Following the merger, I started to receive hate mail and anonymous phone calls. These letters, in the main, were appearing under the door in my office whilst the phone calls were mainly to my home. The mildest content of these letters constituted "Get out you black shit" but the phone calls had more sinister messages in which Priya was often threatened. The Police would regularly camp outside our home overnight. Finally they suggested that a camera should be installed outside my office and the hate mail mysteriously stopped.

The culprits were never caught but students had been ruled out in the investigation.

Simultaneously the new administration, for reasons unbeknown to me, seemed to view me as a nuisance and tried to hinder me professionally. I was given a heavy teaching load more akin to a Junior Lecturer (I was a Senior Professor) and any applications for research facilities were routinely blocked. The University also avoided giving me any recognition for starting the Astronomy Research Programme which I had instituted in 1973 and was now thriving. Notably, when the Queen was invited in 1993 to open the new Astronomy Building (Queen's Building on Newport Road), I was conspicuously left out from the list of invitees.

The above are just a couple of incidents in what really did seem like a campaign against me and continuing right up to the present day. Their discrimination against me was the subject of a Home Office Enquiry in 1996 and at that time the University made a few "without prejudice" concessions such as allowing my research to be conducted under a new Cardiff Centre for Astrobiology. But not much had really changed and I now had to contend with the fall-out from taking them to task.

It was fortunate for me that I was to have a brief respite from this brewing turmoil by virtue of an arrangement that had been made with the old University administration, before the merger. I was granted leave of absence that I took up in two parts during the academic year 1993/1994. One part was in Kingston, Jamaica where I was a Visiting Professor in the Physics Department at the Mona Campus of the University of the West Indies, and the other in the Institute of Space and Astronautical Studies in Tokyo. My setting off on these trips was not without protest, however. The Head of the School of Mathematics, Geoff Griffiths, tried his level best to stop the leave. His argument was that other members of his department who had not taken as much leave as I had done must be given priority. He did not succeed and I was able to take the leave of absence that had been promised by an earlier administration.

Before setting out to take up these assignments I had a pleasant surprise — a surprise that was particularly welcome in the midst of my problems in the Cardiff scene. I received a message that the President of Sri Lanka was to confer a national titular honour *Vidya Jyothi* (equivalent to Knight of Science). Since I have always treasured my links with Sri Lanka I was overwhelmed with delight. Priya and I set off yet again to our native Sri Lanka to the formal investiture ceremony at President's House, Colombo on 22nd May 1992 to be conferred this honour. This Sri Lankan national honour was, I believe, given to me in recognition of my contributions to the cause of science in Sri Lanka, and in particular for my role in setting up the Institute of Fundamental Studies.

On our many visits to the United States over the years Priya and I had stayed with Phil and Sheila Solomon and enjoyed their hospitality in their home in Long Island Sands. Phil was the first mainstream astronomer who had openly and fearlessly supported our cosmic theory of life and the arguments for it at a very early stage. Phil's wife Sheila was a famous sculptor and had often told us of her lifelong ambition to make a bronze statue of Fred Hoyle whom she admired greatly, and her ambition was ultimately realised. Following many reluctant sittings on the part of Fred the statue was finally made, cast in bronze, and given to the University of Cambridge. In July 1992 there was a memorable inauguration event to celebrate the dedication of the statue at the Institute of Astronomy. Many of Fred's collaborators and former students gathered there to reminisce and pay their tributes to the great man. Speeches by John Faulkner, Herman Bondi, Margaret Burbidge, Tommy Gold and Jayant Narlikar all extolled Fred's unique qualities both as a scientist and a friend. Priya and our children were among the invited guests and this was yet another very special moment in my life — revisiting the University and Institute that made me what I am.

In December 1992 Priya, Janaki and I went to the Canary Islands for a two week Winter School of Astrophysics in Tenerife. I was one of nine lecturers at this Winter School. Here I met astronomer Antonio Mampasso who was doing observational work on interstellar dust. I collaborated with him on a joint paper involving the interpretation of his data on the basis of bacterial dust models. At this meeting I delivered a public lecture in the City Hall to a packed audience entitled "The Origin of Life — An astronomer's view". The lecture was simultaneously translated into Spanish and was exceedingly well received.

After all these unscheduled diversions to my programme, my sabbatical year finally began. In 1993 I spent a term working at the Institute of Space and Aeronautical Studies in Tokyo, and in 1994 another term as a Visiting Professor in the Mona Campus of the University of the West Indies in Kingston, Jamaica. My wife Priya and children accompanied me, and on each occasion we set up temporary homes in these vastly different locations.

Our Japanese sojourn led also to the start of a friendship and collaboration that was well outside the bounds of astronomy — my link with Daisaku Ikeda began. Some years earlier a London Publisher, McDonand, approached me to ask if I would write an extended Foreword to a book by

Daisaku Ikeda entitled "Unlocking the Mysteries of Birth and Death: Buddhism in the Contemporary World". (Ikeda is a famous Japanese Buddhist scholar and President of Soka Gakkai International, a vast international lay Buddhist organisation, whose aim is to spread Buddhist values throughout the world.). The content of Ikeda's book had a deep resonance with my own belief that Buddhism and Science are intimately linked. My foreword acknowledged this link and concluded as follows:

> "Since the modern scientific viewpoint is derived from the application of the methods of empirical science, one might wonder how the same results could be reached without recourse to such methods some 2500 years ago. The answer must lie in the still mysterious and unproven powers of meditation. If we are all creatures of the Cosmos would it not seem reasonable that we have an innate knowledge of its nature somewhere deep within ourselves? The basic facts relating to the cosmos could be viewed as essential components of our own true nature....."

President Ikeda was so pleased with my Foreword that he invited me to engage in a dialogue with him for publication in book form. He had published earlier dialogues with eminent scholars including Historian Arnold Toynbee and Chemist Linus Pauling, and so I did not hesitate to accept his invitation.

The result was that between 1993 and 1996 I had several meetings with Ikeda in Japan in order to conduct our dialogue. During these visits the hospitality he lavished on me and my family was most gracious and memorable. We had a unique experience of Japan that left in us a lasting affection for that country and its people. Just as in my native Sri Lanka, Buddhism is very strong in Japan, and the many magnificent temples in Kyoto and elsewhere bear testimony to its far-reaching influence. My dialogue with Ikeda, which was published first in Japanese and later in English, embraced a wide range of topics from philosophy to astronomy. In a reply to one question by Ikeda, I said:

> "The transformation from a Cartesian to a holistic worldview is not easy to achieve for the reason that the former more or less pervades all the institutions of the Western world. The expansion of the European colonisation of

the East in the eighteenth and nineteenth centuries nearly extinguished the indigenous holistic worldview of the Orient. It is only now that societies in the East are beginning to rediscover and reassert their traditional holistic worldview.

A shift towards a holistic paradigm is needed not only in physics and biology, but also in the social sciences — in politics, for instance. We live today on an Earth where the individual components are all ultimately interconnected. A sensible model of Earth must surely include a firm ecological perspective. Such a view is certainly gaining ground, and one could interpret this trend as a move towards a Buddhistic worldview."

Despite our hectic schedule of international travel our home base throughout this time continued to be firmly based in Cardiff and its University. This was where we had set down our roots in 1973 and despite all its drawbacks and the hostilities we encountered, continues to be our home. All our family now live in Cardiff and I love this city with a passion that may be thought unwise. Throughout the 1980's and 1990's Fred Hoyle continued to come to visit us in Cardiff on a regular basis. Needless to say I received much encouragement from his visits, particularly to feel that we were confronting the really important problems of the Universe, against the backdrop of which my own petty squabbles with the University could be viewed in proper perspective.

By the early 1990's a conceptual framework for a grand theory of cosmic life was fully in place, and its predictions were being borne out in observations from several disciplines. Interstellar dust and cemetery dust were found to possess exactly the properties we had predicted, and the oldest life on the Earth was pushed back to a time when intense cometary bombardment was known to have been taking place. We had the strongest indication that comets seeded the planet with life some four billion years ago. The discoveries of microbial life enduring the most extreme conditions were suggesting an alien context for all such properties, and opening possibilities of microbial habitats in a wide range of bodies in the solar system. A wrong theory does not come up repeatedly with such an amazing series of successes. Sooner or later a contradiction turns up and the theory has to be abandoned. This has not happened in our case. Why then is there such deep-rooted hostility to these ideas? The simplest answer was that such ideas went against the grain of an essentially geocentric scientific culture.

There were financial considerations as well. Our work on the cosmic theory of life did not attract substantial grants. A golden age when philosophers and academics held sway had given way to an age of hard headed accountants. Money was all that mattered. The search for truth was subjugated by an overpowering greed for accumulating research funds and political power. Nor were these unsavoury developments confined to the cloisters of academia. Indeed it appeared that Universities were merely responding to major changes that were taking place in the world at large. Margaret Thatcher, whose government introduced the idea of a market economy for universities, was now in her third term of office as Prime Minister. People were reckoned to be less important than dubiously conceived objectives within institutions, including Universities.

Intolerance of all sorts was on the rise. In 1989 Salman Rushdie published his novel *The Satanic Verses*, and the Iranian leader Ayatollah Khomeini orders Muslims to execute him. In Sri Lanka conflicts between Tamil separatists and the Government flared up sporadically. In China racial attacks on black students early in 1989 were followed by the historic protest and massacre of dissidents in Tiananmen Square. Even in the UK the incidence of racial attacks were noticeably on the increase, with the Police and authorities often turning a blind eye. My wife and I continued to have our fair share of "go home Paki" comments in Cardiff, and similar written threats were delivered to me at the University by persons who could not be identified.

Despite all this I continued to pursue my researches doggedly in whichever direction that new data directed. And new data did indeed come our way at a brisk pace. A discovery of a 3.28 micrometre emission feature in the infrared in the diffuse radiation emitted by the Galaxy confirmed that aromatic molecules of some kind were exceedingly common on a galactic scale. We argued that the infrared emissions not just at 3.28 micrometres but over discrete set of wavelengths — 3.28, 6.2, 7.7, 8.6, 11.3 micrometres — must arise from the absorption of ultraviolet starlight by the same molecular system that degrades this energy into the infrared. We had shown much earlier that the 2175Å extinction of starlight may be due to biological aromatic molecules, and it seemed natural then to connect the two phenomena.

Thus we developed a unified theory of infrared emission and ultraviolet extinction by the same ensemble of aromatic molecules. A currently fashionable non-biological aromatic molecule was coronene ($C_{24}H_{12}$) and it was easy to demonstrate that this type of molecule was nowhere near as good as biological aromatics.[31]

Fred Hoyle and I both attended the 22nd ESLAB Symposium that was held in the delightful Spanish town of Salamanca between 7 and 9 December 1988. We presented three joint papers here, two on our theory of interstellar grains and one on a model of the cosmic microwave background based on long iron whiskers. Apart from the presence of the predictably hostile Greenberg contingent, we felt there was a mellowing of attitude towards us, compared to our experiences both in Colombo and at the RAS a few years earlier. There was still a long way yet to conceding a biological grain model, but people were at least willing to listen to the arguments and to extract what they thought was its acceptable essence. Organic dust everywhere in the cosmos, including in the comets, had become the order of the day.

Quite a different kind of development that attracted our attention in the winter of 1989 was an outbreak of influenza throughout the UK. It was described as the worst influenza epidemic for 12 years, and hospitals were inundated with cases arising from complications of the flu. Fred Hoyle and I decided to undertake our second Hoyle–Wickramasinghe Influenza Survey by sending out questionnaires to all independent schools in the UK, and by visiting particular schools in our immediate neighbourhood. Priya and I went in person to discover what happened in a place not far from Cardiff. We discovered that among the earliest to succumb in the new outbreak were the inhabitants of the sleepy little village of Gowerton, near Swansea. Attendance registers at the local school showed a sudden rise of absences on 27 November, the same day when a local publican, his wife and their child all came down with flu. Furthermore it was documented that the entire village became shrouded in a persistent low-lying mist starting about two days before the outbreak.

Although the influenza type known as H3N2 was the dominant subtype found to be involved in the epidemic, a whole host of other airborne viruses

---

[31]F. Hoyle and N.C. Wickramasinghe, *Astrophys. Sp. Sci.*, 154, 143–147, 1989; N.C. Wickramasinghe, F. Hoyle and T. Al-Jubory, *Astrophys. Sp. Sci.*, 158, 135–140, 1989.

appears to have been in circulation at the same time. The incidence patterns of these viruses were consistent with an atmospheric fall-out model, and inconsistent with direct person-to-person spread. All the data we collected appeared to corroborate our findings from the 1977/78 pandemic — even the epidemic patterns at Eton College. At this stage we were surprised to receive an invitation from the Royal Society of Medicine to write a review paper of our findings about influenza. Our article appeared as: "Influenza — evidence against contagion: discussion paper".[32]

---

[32] F. Hoyle and N.C. Wickramasinghe, 1990. Evidence for non-transmissibility of influenza, *J. Roy. Soc. Med.*, 83, 258.

CHAPTER 19

ICE AGES AND THE GENESIS OF RACISM

On the journey of life one comes across an occasional traveller with whom a bond of kinship is immediately struck. One such individual Brig Klyce came into my life in the late 1980's with an offering of a 200 page thesis on which I was asked to comment and possibly help get published. In his manuscript he presented the case for our cosmic ancestry in as compelling a manner as it was possible to make at the time. The consonance with my own point of view, that I have described thus far, was so close that it was inevitable that we struck up a correspondence and eventually a working relationship and friendship that still continues.

The fate of Brig Klyce's original manuscript was as would have been expected for any radical challenge of an orthodox position in science, particularly a challenge coming from one who had no formal "credentials" in the field. Brig's original manuscript eventually became the framework for what is now the best available internet site for panspermia research — a news source, a research tool for scholars as well as a discussion forum that contributes to the ongoing debate.

Besides the "Cosmic Ancestry Website" Brig Klyce started the "Astrobiology Research Trust" of which he is the Chairman. He has supported many research projects related to panspermia and has stepped in on many occasions to fund our research activities in Cardiff at times when everyone else had turned their backs. Brig and I continue to collaborate on the most exciting of all research ventures at the present time — unravelling our cosmic ancestry.

In the summer of 1989 Jayant Narlikar contacted me to ask if I might be able to host a small cosmology workshop in which Fred Hoyle could discuss the current state of cosmology with his closest collaborators. We first considered Gregynog Hall in mid-Wales as a possible venue for the meeting but decided against it because of the logistics of transporting participants to and from airports. (Gegynog Hall is over a 100 miles distant from an international airport, and roads to it are not the best). Between 25 and 29 September 1989 Fred Hoyle, Geoffrey Burbidge, Harlton Arp, Jayant Narlikar and I met at a small residential conference centre — Dyffryn Gardens — just outside Cardiff city. We discussed the many lines of evidence that all appeared to go against the standard Big-Bang model of the universe.

Our deliberations at the meeting were written up in the form of an article and sent to John Maddox, Editor of *Nature*. It is still a little bewildering how Maddox was persuaded to publish such a devastating attack on conventional cosmology. The paper appeared with the title "The Extragalactic Universe: an alternative view" in the issue of *Nature* of 30 August 1990.[33] As time has gone by cracks in the standard Big Bang cosmologies have come to the fore resulting from the deployment of a new generation of more powerful telescopes. Eternally "inflating" and cyclic universe models are under discussion, although cracks in the popular theory continue to be skilfully papered over. The situation in cosmology is analogous to the epicycles of Ptolemy. It is amply clear now that the last word on these matters has yet not been said.

In 1996 Hoyle and I collaborated with Bill Napier and Victor Clube to explore the effects of the breakup of a giant comet, leading to fragments in Earth-crossing orbits, and to recurrent episodes of impacts and cometary dusting on the Earth. We argued for a connection between such cometary events, periodic glaciations as well as episodes of mass-extinctions in the geological record. We also suggested that the entire history of human civilization, over the past 10,000 years, after the end of the last ice age, bears witness to a record of repeated episodes of assaults from the skies.[34] In a later paper we also worked out a more specific analysis of the possible connection between cometary events and ice ages.[35] These ideas are also receiving further support as time progresses.

---

[33] H.C. Arp, G. Burbidge, F. Hoyle and N.C. Wickramasinghe, 1990. *Nature*, 346, 807–812.
[34] S.V.M. Clube, F. Hoyle, W.M. Napier and N.C. Wickramasinghe, 1996. *Astrophys. Sp. Sci.*, 245, 43–60.
[35] F. Hoyle and N.C. Wickramasinghe, 2001. *Astrophys. Sp. Sci.*, 275, 367–376.

Our modelling of ice ages in the 1990's took a somewhat unexpected turn touching as it did on a highly topical social issue. The connection might sound bizarre, but our studies of ice ages led us to model of the origin of race prejudice among humans. This issue had come recently to the fore in view of a much publicised report of a British government inquiry into the gratuitous murder of Stephen Lawrence (a black youngster) in South East London, in April 1993. The London Metropolitan Police had refused to prosecute the white youths who murdered Lawrence, and the family launched a private prosecution that led to a Government inquiry under the chairmanship of Sir William Macpherson. The Macpherson report, published in March 1999, found the Metropolitan Police to be "institutionally racist", a phenomenon that was said to exist also in other institutions.

I wrote the following letter to the London *Independent* on 26 February 1999, in the wake of the McPhearson report on the Lawrence murder:

"Sir — Rooting out of racism from our society will remain an improbable Utopia unless our educational system can be drastically overhauled.

It has often been stated that education is the key to changing social attitudes towards greater tolerance and compassion, but there is little evidence that such a polemic is put into practice. Our children ought to be imbued with trans-cultural values and encouraged, from an early age, to appreciate that non-Caucasian races have made significant contributions throughout history in all aspects of human endeavour, ranging from mathematics and astronomy, through philosophy to the visual and literary arts. A history of mathematics that does not accord the Indians their due place in the subject (the Indians invented the concept of zero, for instance) is racist beyond dispute.

My own experiences as an Asian academic and parent living and working in Britain for over three decades have convinced me that racism is rife in our educational system. With frequent instances of children being subjected to "paki" insults, and academics covertly mistreating their black and Asian counterparts, it is clear to me that change is long overdue. There can be no future for a multicultural Britain unless our highest seats of learning set an example to the rest.

CHANDRA WICKRAMASINGHE
Professor of Applied Mathematics and Astronomy
University of Wales, Cardiff "

This letter was written some 15 years ago, but my beliefs have not changed with the passage of time.

How, one might ask, could such a strong emotional response to such a trivial difference in skin colour prevail at a time in our civilization when we pride ourselves as being "enlightened"? When Fred Hoyle and I first discussed this, we soon agreed that there must be a powerful biological imperative for racism to persist, not just in Britain, but throughout much of the modern world. We published our speculations on this subject in 1999 in the Journal of Scientific Exploration.[36]

It is an undeniable fact that human evolution over the past two million years has led to the emergence of two broadly distinct groups of people with regard to skin colour, one fair the other dark. The lighter skinned group now occupies countries in northern latitudes that were on the borders of glaciers during ice ages, and the darker skinned group mainly inhabit temperate and equatorial regions. The difference in skin colour between these groups hinges on a variable efficiency to produce the pigment melanin. Melanin is produced in a special group of cells known as the melanocytes that are located at the base of the skin, and although the density of such cells remains more or less invariable, the efficiency of melanin expression is highly variable. Many genes appear to be involved in melanin production, and the overall situation for melanin expression (that is to say for being black) is strongly dominant.

The present-day situation for maintaining selective pressures for melanin expression rests on a razor's edge between two competing effects. On the one hand melanin as a pigment that protects the base of the skin from being damaged by ultraviolet radiation from the Sun which in extreme instances leads to carcinomas of the skin. On the other hand, an adequate penetration through the skin of ultraviolet radiation with wavelength shortward of 3130Å is needed for the production of vitamin D. Whilst the latter requirement is not too relevant in the present day with high levels of

---

[36] F. Hoyle and N.C. Wickramasinghe, 1999. *J. Sci. Exploration*, 13, No. 3, 681–684.

nutrition generally, it would have been a strong selective factor for survival in harsher prehistoric times. With lower levels of dietary acquisition of vitamin D, the lack of an adequate absorption of sunlight would lead to the crippling disease of rickets. In this disease severe bone deformities result from the lack of vitamin D, a substance that plays a crucial role in the absorption of calcium from food. The correct level of pigment expression depends on the available ultraviolet light at any given location on the Earth at a given time. The balance is between rickets causing skeletal deformities with a consequent lower fecundity, and excessive sunburn radiation leading to skin cancers with attendant high levels of mortality.

The higher incidence of skin cancers in lighter skinned Caucasian migrants in the tropics is well attested. Likewise a high incidence of rickets has been recorded in black and Asian populations living in northern latitude countries before the large-scale introduction of vitamin supplements into staple foods. At the beginning of the 20th century, rickets was reported to affect 90% of black infants in New York. Even as recently as the 1970's high rates of incidence of rickets have been recorded in children of Asian immigrants living in Britain, as for instance in a Glasgow-based study. The intensity of sunlight available to Asian immigrants in countries like the UK is obviously mismatched to their expressed pigment level, but routine food fortifications and vitamin pills now generally compensate for this deficiency. Needless to say such dietary supplements were unavailable in prehistoric times.

Throughout the Pleistocene epoch as human evolution progressed, the Earth was locked in an ice age that lasted for nearly 2.5 million years. There were warmer interglacial remissions interspersed throughout this time, each one lasting for about 10,000 years, and the total duration of all such warm periods made up just 10 per cent of the entire Pleistocene era. The Earth emerged from the last ice age approximately 11,000 years ago.

During ice ages the average temperature of the Earth's surface was about 10 degrees Celsius colder than today, and ice sheets were about three times as extensive as they are now. White skinned Nordic tribes living close to the edge of rugged windswept ice sheets under grey skies would have been eking out a precarious existence, grabbing whatever food could be gathered and utilising every photon of ultraviolet from the sun in order to stay alive and free of rickets. For them survival was crucially contingent upon having their genes for melanin suppressed. For people living in the tropics, however, the drier ice-age conditions with less cloud cover than at present would have made for a remorseless flood of ultraviolet radiation to fall on their skins. Survival for them was contingent on the fullest expression of their melanin genes, being as black they possibly could be.

In the course of random migrations from the south, the white populations living at the edge of ice sheets would have been at risk through mating with people possessing darker skins. Black-white mating would have tended to produce offspring with darker skins and thus more prone to rickets. Fewer of these malformed children would reach reproductive age, so black-white mating posed a real extinction threat to the white races. Under such circumstances the emergence of mating prohibitions and colour prejudice would be a natural outcome. The prejudice would become deeply ingrained in social traditions, language, mythology and religion. Thus the depiction of Satan as a black figure cannot be regarded as accidental, nor can the association of evil generally with blackness in Western traditions. The strong emotions manifest in modern racism could perhaps be understood, but not forgiven, in these terms. It is my belief that some, though by no means all, the opposition I faced in advancing our ideas on the cosmic origins of life may also be connected with the same prejudice. Could a predominantly Anglo Saxon scientific community take lightly the prospect of having their most cherished scientific paradigm overturned by an outsider to that community?

One of the most significant conferences to which I was invited was an IAU colloquium on Bioastronomy (IAU Colloquium No. 161). This was held in the delightful island of Capri in July 1996 and Priya and I simply could not pass over this invitation. The main conference convenor Cristiano Cosmovici later told me that my invitation was opposed by certain United States scientists. But as a good Italian with the experience of Galileo and Bruno behind him he felt it important to allow my point of view to be expressed and defended. This being the only way that science can be freed from the shackles of conservatism. My talk entitled "Infrared signatures of prebiology — or biology" addressed a question that nearly two decades later still remains highly charged. My answer to this question did not receive the approbation of the majority attending the meeting. It was insisted that the idea of life disperse in interstellar dust clouds was in Carl Sagan's words "an extraordinary claim and extraordinary claims need extraordinary evidence to defend them." This polemical rebuttal of evidence can be turned on its head by asserting that the claim of life confined to Earth is by far the more extraordinary claim. But the conference tended to rejecting hard evidence in favour of the cosy comfort of conservatism. There were, however, high-points of the meeting that I enjoyed — discussions relating to the comet Shoemaker-Levy 9 that broke up into a "string of pearls" that collided with Jupiter in 1994, and the first confirmed planet outside our solar system orbiting the 51 Pegasi in the constellation of Pegasus.

A few memorable plenary talks were given, including lectures by Nobel laureates Christian de Duve and Charles Townes. Christian de Duve (the discover of lysosomes within cells) talked about the subject of his new book *Vital Dust — Life as a Cosmic Imperative*, which I found rivetingly interesting. Charles Townes, recognised for his discoveries relating to masers, talked about such devices in connection with the search for extra-terrestrial intelligence (SETI). I also had the opportunity to meet Frank Drake, the unquestioned pioneer of SETI — a program to search the skies for extraterrestrial civilisations — that still limps along with the barest of public funding. These contacts alone made the trip worthwhile. Above all it was a fabulous experience for Priya and me to enjoy an exquisite conference venue and the most beautiful island of Capri.

Returning to the timeline of my personal research, an important development was the discovery of a new ally, Milton Wainwright of Sheffield University. Milton entered my story serendipitously in 1996 after he had published a short piece in a microbiology journal in defence of panspermia. This was followed by a correspondence in which he offered assistance in any way that could lead to further progress towards the acceptance of panspermia. Milton's experience as a microbiologist combined with his abiding interest in the history of science placed him in good stead to contribute to the debate concerning panspermia.

CHAPTER 20

A NEW MILLENIUM AND THE TASTE OF THINGS TO COME

Besides still vigorously pursuing my scientific goals, the first decade of the 21st century heralded for me and Priya a new type of adventure. We spent a few weeks every year sailing to interesting and exotic places aboard cruise ships on which I would give lectures on astronomy and Priya sometimes gave demonstrations and talks on Indian and Sri Lankan cooking. After the voyages I had enjoyed in earlier years, these brief spells at sea were both a reminder of the past and as well as a welcome respite from the cares of a turbulent world.

Among our many cruises, one to the Galapagos Islands in 2009 was the most memorable and also most significant to my present narrative. These volcanic islands of enormously variable climate and terrain had a profound relevance to the formulation of Darwin's theory of Natural Selection. When Darwin set out on his famous voyage on HMS Beagle in 1831 there was evidently no prior plan to set anchor off the Galapagos Islands. After collecting specimens from South America it is said that they stopped in these islands merely to collect tortoises for meat for their continuing voyage.

What had struck Darwin with great force, as indeed it struck me in 2009, was that the animals inhabiting these islands — e.g. Darwin's finches — were in many cases unique. Furthermore, they differed from related species on the mainland and also differed from one island to the next. Darwin inferred that in these islands that were separated from the South American continent by more than 500 miles, species came to be adapted according to diverse local conditions. Indeed we noticed the islands differed so dramatically one from another that adaptation was almost mandatory for survival. To this extent I agree with Darwin's conclusion but the mechanism by which all the required changes are wrought is by no means fully understood on the basis of an Earth-bound theory of life. The explanation of the facts relating to the animals in the Galapagos Islands must extend beyond the accumulation of small mutations, in my view. According the cosmic theory of life an almost infinite diversity of genes exists in a cosmically generated gene pool that continually interacts with terrestrial environments. Local niches mainly serve to select those cosmic genes from an almost "infinite set" that are best suited for the survival of any particular species. A highlight of our visit was to catch a glimpse of the giant tortoise "lonesome George", an animal which Darwin himself would have seen.

In the year 2000 the formation of the Cardiff Centre for Astrobiology freed me from the control and jurisdiction of the University's School of Mathematics that had been so traumatic of late. This Centre, the first of its kind in the world, came directly under the Vice Chancellor to whom I was nominally responsible. A complication, however, related to my research students, who had to come from some academic department, so I had to reluctantly maintain a weak link with the School of Mathematics.

Fred Hoyle and Bill Napier continued as Honorary Professors, and Max Wallis and Shirwan Al-Mufti were Honorary Research Fellows at the new Centre. Shortly afterwards Milton Wainwright also joined our ranks as an Honorary Professor. The new centre gradually began to acquire an international status as the first of its kind in astrobiology, and a reputation for being the only research Centre actively engaged in Panspermia research. Our first major research undertaking in the new millennium was a collaboration with Indian scientists under the leadership of Jayant Narlikar and the Indian Space Research Organisation. The aim was to detect bacteria entering the Earth by examining samples recovered from the high stratosphere.

The laboratory work in Cardiff was carried out in the School of Biosciences under the guidance of David Lloyd, Professor of Microbiology at the University and with the assistance of research student Melanie Harris. The first phase of this investigation was completed in July 2001 and we obtained unambiguous evidence for the presence of clumps of living cells in air samples from as high as 41 kilometres, well above the local tropopause (16 km), above which no micron-sized aerosols from lower down would normally be expected to be transported. The detection was made using electron microscope images, and by application of a fluorescent dye known as carbocyanine that is only taken up by the membranes of living cells.

The variation with height of the density of such cells indicated strongly that the clumps of bacterial cells were falling from space. The input of such biological material was provisionally estimated to be between 1/3–1 tonne per day over the entire planet. If this amount of organic material was in the form of bacteria, the annual input of bacteria is a staggering 10^{21} in number amounting to about a tenth of a tonne averaged over the whole Earth. These results were presented by me in July 2001 on behalf of our team at the *Instruments, Methods, ad Missions for Astrobiology IV* session of the *SPIE* Meeting in San Diego.[37] The presentation made international headlines. Doubts about contamination were naturally raised by sceptics imbued with a geocentric worldview, but our initial results have since received extensive confirmation in later work including recent studies by a group of scientists in Japan.

While the work carried out in Cardiff failed to grow microbes from the stratospheric collections, Milton Wainwright[38] succeeded in culturing two separate bacterial species that were related to known terrestrial species. Although this work was published in a reputable microbiology journal that was subject to rigorous peer review, it should be placed on record that hostile comments sprang from many sources. One comment stating that the cultured microbes, despite all the controls that were used, had to be contaminants for the reason that they were similar to terrestrial species shows that our critics have not read our thesis regarding "evolution from space". If microbes on earth are derived from comets, and continue to be replenished on the timescale of millions of year, then it is to be expected that new organisms will be similar to ones that are resident on Earth. Minor evolutionary genetic drifts are all that are expected, and these are indeed what are found.

---

[37] Melanie J. Harris, N.C. Wickramasinghe, David Lloyd, J.V. Narlikar, P. Rajaratnam, M.P. Turner, S. Al-Mufti, M.K. Wallis, S. Ramadurai and F. Hoyle, 2002. *Proc. SPIE*, 4495, 192.

[38] Wainwright, M., Wickramasinghe, N.C., Narlikar, J.V. *et al.*, 2003. *FEMS Microbiol.Lett*, 218, 161.

Another project we were drawn into somewhat unexpectedly was a collaboration with the Indian scientist Godfrey Louis to investigate a remarkable phenomenon — the Red Rain of Kerala. In the summer of 2001 red rain fell in large quantity over much of the state of Kerala in south India, and Godfrey Louis, a physicist at Mahatma Gandhi University, conducted a series of laboratory studies on samples of the rain that he had collected.[39] His initial results showed that the red colour of the rain was due to the presence of pigmented biological cells. They were generally similar in appearance to algal cells but they were unlike any algal cell known thus far. Godfrey Louis claimed that the cells could be cultured when they were placed in a hydrocarbon-rich medium under high pressure at a temperature of 450 °C — higher than the survival temperature of any known bacterial or algal cell. The fall of the Kerala rain had been preceded by a loud sonic boom, and it was conjectured that a small fragment of a comet exploded in the stratosphere, unleashing vast quantities of red cells that became the nuclei of raindrops. This intriguing story had an obvious possible link with cometary panspermia theories, so it was natural for Godfrey Louis to seek a collaboration with us. A sample of the rain was sent to Cardiff and my colleagues and I have worked on this for nearly a whole decade. It may sound surprising, but it remains the case that to this day we have not been able to identify these cells. The biological nature of the red cell material was confirmed by Milton Wainwright and a PhD student Gangappa Rajkumar, and also its replication under high pressure to a temperature of 121 °C.

I shall now digress from my scientific story to record the passing of two individuals who were profoundly important in my life. First and foremost my mentor, friend and long-term collaborator Fred Hoyle passed away in August 2001. I was invited to write his obituary for the *Independent* (23 August 2001), in which I said:

"Throughout a long and distinguished career stretching over six decades Fred Hoyle sought to answer some of the biggest questions in science.

---

[39] Godfrey Louis, A. Santhosh Kumar, 2006. The red rain phenomenon of Kerala and its possible extraterrestrial origin, *Astrophys. Sp. Sci.*, 302, 175–187.

How did the Universe originate? How did life begin? What are the eventual fates of planets, stars and galaxies? More often than not he discovered answers to such questions in the most unsuspected places. Hoyle believed that, as a general rule, solutions to major unsolved problems had to be sought by exploring radical hypotheses, whilst at the same time not deviating from well-attested scientific tools and methods.......

Hoyle's work on nucleosynthesis in collaboration with William A. Fowler and Geoffrey and Margaret Burbidge led to our present understanding of the origin of chemical elements in stars...."

In a personal letter to Barbara Hoyle I struck a more personal note:

"Words cannot express my emotions on this saddest of all occasions but I shall try.

My thoughts and feelings are with you. We mourn together the death of a truly great man, you a devoted husband, me a friend, teacher and mentor. Perhaps we can take comfort from one indisputable fact: his genius adorned the intellectual world of the 20th century, and the world will surely be the poorer for his passing. Priya joins me in extending our condolences to you, and hope that you may have the fortitude to bear this terrible loss.

I shall personally miss dear Fred at several levels. First and foremost as a friend to whom I owed my entire career and development as a scientist, and whose advice I valued and treasured at all times. Even upto a few weeks ago it was a great comfort to know that he was there at the end of a telephone line, and now suddenly he is no more.

I recall vividly my first meetings with Fred at 1 Clarkson Close, an intrepid student from a distant colony, lonely and unhappy, and how you, dear Barbara, welcomed me so affectionately into your family.

As my working relationship with Fred matured and developed over the years I came to rely heavily on his wisdom and judgment in the course of the ideas I pursued. I consider it the greatest privilege of my life to have been able to join him in exploring perhaps the most important problem in the whole of science — how we came into being. If I had led Fred along inappropriate paths, as some commentators have said, I am sorry, but we had both agreed to pursue unrelentingly the path towards the Truth. We may have made enemies in the process but I am sure that was worth it in relation to the magnificence of the vistas of thought that were opened.

Fred was acknowledged and acclaimed by the world of science for taking a small but profoundly important step towards understanding our origins, the stellar synthesis of the chemical elements of which we are made. But I am confident that his much bigger, more unorthodox leaps of thought relating to the origins of life will also prove right. The decisive evidence is just round the corner, and I regret that Fred will not be there to see it. In the fullness of time, when the dust has settled on all the controversies and disputes that were raised by this work, I am sure that Fred Hoyle would share a place with Galileo, Copernicus and Newton in the annals of history.

My memories flit now between the happy times we spent at 1 Clarkson Close in the company of the Hoyles, the many occasions when you entertained me, and the many times that Fred visited Cardiff in the 1980's and 1990's. His brilliant lectures to packed audiences, our time together at a conference in Sri Lanka, his stimulating company at dinner......These occasions were the highlight of my life, and I shall treasure their memory for the rest of my days, however long or short they might be."

Four years later in August 2005 my father, who had also influenced me much earlier in career, also passed away, and I wrote his obituary for the *Ceylon Daily News*:

"....I believe it was his example, shining like a beacon, that guided us to where we eventually reached. I myself owe my intellectual development to this great and gentle man — a man who had an infectious passion for scholarship, mathematics and the English language — poetry in particular. Long before I had come to learn higher mathematics at school I vividly remember how my father introduced me to "tricks" using Calculus to solve elementary problems. My love for mathematics grew from these encounters. He showed me both the elegance of mathematical logic and the role it could play in understanding the Universe...."

On a happier note, in this decade my daughter Janaki, who obtained a first class honours degree in mathematics from Bath University, joined as a research student at the Cardiff Centre for Astrobiology and completed a PhD under the supervision of Bill Napier. Janaki and Bill Napier performed detailed calculations of the dynamical effects of close encounters of our solar system with massive molecular clouds, and showed that they resulted in episodic increases in the rates of comets' impacts onto the Earth. Such impacts lead to an inevitable splash-back into space of material from Earth that contained viable biological cells and genetic material from evolved life forms. These genetic materials from Earth can then reach newly-forming planetary systems, seed life on alien planets, contribute to the evolution and give an added dimension to panspermia. Darwinian evolution can no longer be considered to be confined to Earth — it must occur on a truly galactic scale — The products of evolution in one galactic location being intermixed with those of another.

I also became involved in hosting two international conferences. The first from 24–26 June 2002 was entitled "Fred Hoyle's Universe" to celebrate Fred Hoyle's extraordinary contributions to science. We had lectures from leading figures in the world of astronomy, who had connections with Fred Hoyle during his long career. They included Hermann Bondi, Geoffrey and Margaret Burbidge, Don Clayton, John Faulkner, Leon Mestel, Sverre Aarseth, Chip Arp, Phil Solomon, Jayant Narlikar, John Maddox and Martin Rees. We also had a live video link with Arthur C. Clarke in Colombo. This was indeed a most memorable occasion, and one that helped to place the Cardiff Centre for Astrobiology on the international map. The second conference was on "Cosmic Dust and Panspermia" on 5–8 September 2006, which was intended to mark my formal retirement from the full-time tenured Professorship to which I was appointed in 1973. Again this conference attracted a large number of prominent scientists including the Burbidges, Richard Hoover, Phil Solomon and Gil Levin and it attracted much media attention as well. Both our international conferences in this decade were supported by generous grants from Brig Klyces's Astrobiology Research Trust.

On the 26th of December 2004 a massive tsunami struck the coast of Sri Lanka and caused death and destruction on an unprecedented scale. Vast tracts of beautiful coastline were ravaged; some 30,000 people were killed and a half million made homeless. It took the country well over three years to recover from the shock of this terrible disaster and to rebuild the damage. Soon after the Tsunami many international organisations came to help in the rebuilding efforts but progress towards returning to normality was slow. Priya began to help in her own way by collecting donations from friends and distributing urgently needed books and other resources to schools in the affected areas. All this happened while the civil war between the Tamil Tigers and the Sri Lankan Army continued to rage, with sporadic incidents of terrorism constantly threatening the peace and tranquillity of the island. In May 2009, after a bitterly fought final battle, the Tamil Tigers were at last defeated and peace returned after 25 years of civil war. President Rajapaksa's government now has the difficult task of rebuilding a nation and restoring it to the glory and prosperity of former times.

Back in Cardiff in September 2010 it should be recorded that at the end of a decade of its existence the Cardiff Centre for Astrobiology, a Centre that derived its inspiration from its association with Fred Hoyle, had a premature closure imposed on it by the University. Britain now was facing its worst recession since 1946, the time of my first visit to this country as a child. All publicly funded institutions had to impose stringencies and cut-backs in order to survive, and Universities were no exception. It was said that our research activities did not attract much funding from outside bodies, and to some extent this was true. So it was that the first astrobiology centre in the world ended its short life, not with a bang but a whimper.

By a stroke of luck, however, all the research activities of Cardiff Centre for Astrobiology were moved to the University of Buckingham which has taken over our entire team of distinguished Honorary Professors and Research Fellows. All the work that we were formerly doing under the aegis of the Cardiff Centre for Astrobiology was now transferred to the Buckingham Centre for Astrobiology (BCAB) at the University of Buckingham.

Another development of note was my collaboration with Rudy E. Schild (Harvard-Smithsonian Centre for Astrophysics) and Carl H. Gibson (UCSD) in an innovative venture of starting the new online *Journal of Cosmology*. As joint editors of this journal we are able to overcome a modern tendency to give publication credibility only to those research projects that are considered orthodox in relation to current controversies. I noted earlier that we had experienced difficulties of this kind in the 1980's in connection with the cosmic origins of life. Similar levels of censorship now apply to other areas of science as well, in particular to non-standard cosmologies that are not perceived as conforming to prevailing dogmas. I think this is an exceedingly dangerous trend that needs to be checked if we are not to drift into the type of repressive culture that prevailed throughout the Middle Ages. Our peer-reviewed, open-access online *Journal of Cosmology*, we hope, might in some small way help to overcome this trend and could be a trailblazer for similar developments in the future.

In November 2012 red rain episodes similar to what happened in Kerala in 2001 were reported over large areas of central Sri Lanka. Investigations of the Sri Lankan red rain have shown that the red cells present in this rain are very similar to the unidentified cells of the Kerala red rain that I already discussed. A significant difference, however, was the nearly simultaneous reports of fireball sightings and a fall of carbonaceous meteorites, strongly suggesting a causal link.

It would seem curious that serendipity has intervened in the closing act of this drama by providing evidence from my own native Sri Lanka that carries a promise of fulfilling my "destiny". The Sri Lankan red rain that was analysed so far has led to results that are fully in accord with a likely space origin of the red rain cells. The analysis of the meteorite samples conducted by Jamie Wallis and Daryl Wallis in Cardiff and independently by Richard Hoover at the Marshall Space Flight Center in Housten have produced results that are so amazing that they have provoked intense argument and controversy.[40,41] The Sri Lankan meteorite contains within it

---

[40] J. Wallis, N.C. Wickramasinghe, D.H. Wallis, N. Miyake, M.K. Wallis, R. Hoover, A. Samaranayake, K. Wickramarathne and A. Oldroyd, 2013, *Proc. SPIE*, 8865, 886508.

[41] Richard B. Hoover, Jamie Wallis, K. Wickramarathne, A. Samaranayake, G. Williams, G. Jerman, D.H. Wallis and N.C. Wickramasinghe, 2013. *Proc. SPIE*, 8865, 886506.

fossilised extinct microbes and fossilised diatoms that are beautifully pre-served. These new discoveries clearly confirm that life is a truly cosmic phenomenon. The only way out for the critic is to assert that the "meteor-ites" are not genuine meteorites, but are either rocks from the Earth or they were artificially synthesised. Not only do such assertions fall foul of the well-attested fireball sightings that preceded the fall, but it also contradicts the discovery that the meteorite contains a ratio of oxygen isotopes that are not consistent with Earth rocks and also a high level of the element iridium all of which are consistent with a meteoritic origin. Otherwise we would have to believe that a poor subsistence farmer in Sri Lanka having seen a fireball one night proceeded to manufacture tonnes of synthetic stones which he scattered over several acres of his field on the following morning. Ridiculous may be, but as Julius Caesar said:

"fere libente homines id quod volunt credunt — men readily believe what they want to believe"

In a further development of this saga, my friend Milton Wainwright has flown a balloon borne cometary dust collector to a height of 27 kilo-metres in the stratosphere and brought back microorganisms that were falling from space that are unequivocally alien.[42] All the currently avail-able evidence in 2014 point unerringly to our cosmic ancestry.

At the 2013 Annual Sessions of the Ceylon College of Physicians (organized in conjunction with the Royal College of Physicians) I was the Chief Guest and gave a presentation to this august gathering with the title: "Impact of Discovering Extraterrestrial Life on Humanity". In it I said:

"The discovery of intelligent life outside Earth, if that does indeed tran-spire, poses the most serious problems of all, calling for fundamental revisions and readjustments of our perceptions about ourselves. A widely prevalent obsession with UFO's — for which of course there is no hard evidence — may be an indication of the fear that might be engendered if contact with extraterrestrial intelligence really took place. Even the mere

---

[42] Milton Wainwright, Christopher E. Rose, Alexander J. Baker, Briston, K.J and N. Chandra Wickramasinghe, 2013. *Proc. SPIE*, 8865OL.

proof that such extraterrestrial intelligence exists will seriously erode our perceived position of unrivalled supremacy in the world. And if extraterrestrial intelligence is indeed found to be nearby, and contact thought imminent, the situation might become analogous to the fear that primitive tribes may have had regarding the prospect of encounters with more civilised conquerors."

In August 2013 a new phase in my quest for cosmic origins began. Two new friends enter my life — Gensuke Tokoro, a Harvard-trained Economist and Tokyo Businessman involved in the production of vaccines and Takafumi Matsui, an astronomer at Tokyo University. Gensuke Tokoro corresponded with me earlier in the year requesting a meeting in Cardiff in order to conduct a dialogue on the cosmic origins of life. He had diligently studied the Hoyle–Wickramasinghe thesis of Viruses from Space and convinced his friend Matsui to take the matter seriously. Our meeting and an extended dialogue took place in August 2013 at the Hilton Hotel Cardiff, and this led to the publication of a popular book in Japanese dealing with Cosmic Origins and the Red Rain phenomenon discussed earlier. Gensuke invited me (along with Priya and Janaki) to Japan in the last week of November 2013 to deliver a lecture and to launch the new book. Gensuke graciously and lavishly hosted our visit, entertaining us at the best restaurants in Tokyo and Kyoto.

During this week in November, Gensuke, Matsui and I had many discussion dealing with the question as to how a long-overdue paradigm shift from geocentric to cosmocentric biology could be achieved. The decision was for Gensuke Tokoro to start a new research centre "Institute for the Study of Panspermia and Astroeconomics" (ISPA) with a remit that includes the launching of commercially sponsored balloons into the stratosphere to collect microorganisms from comets on a regular basis, and eventually hopefully convincing the scientific community of our cosmic origins. Whilst Milton Wainwright continues to fly balloons concentrating on the recovery of bacteria and other unicellular organisms, ISPA would attempt to collect viruses and sequence them with a view to determining their origin. The latter is reckoned to be important in view of the fact that viruses are now believed to make up the bulk of the biomass of our planet. A single drop of ocean water has been found to contain some ten million individual virus particles, and the total count of viruses on our planet is estimated at 10^{31}. If these viruses are imagined strung end to end in a straight line the distance spanned would be some 100 million light years. Gensuka Tokoro and I collaborated in a series of publications dealing with such matters,[43] and one paper written with Milton Wainwright was presented at a conference organized by the United Nations Office of Outer Space Affairs in Graz, Switzerland in November 2014. We concluded this presentation with the statement:

"Perhaps the worst casualty of all in admitting this paradigm shift will be our self-image as the supreme lifeform in the cosmos, which will be shattered and require major efforts in psychological and educational readjustment.

Over tens of thousands of years humans have enlarged the size of the unit to which loyalty is owed from the nuclear family, to the extended family, the small tribe, the city state, the nation state and eventually to all mankind. It is likely that the entire structure of human society will undergo a transformation more fundamental and profound than any that has taken place so far. A new cosmic consciousness would dawn."

---

[43]Wickramasinghe, N.C. and Tokoro, G., 2014. Life as a cosmic phenomenon: 1. The socio-economic control of a scientific paradigm, *Journal of Cosmology*, **24**, 12012–12018; Tokoro, G. and Wickramasinghe, N.C., 2014, Life as a cosmic phenomenon 2: the panspermic trajectory of *Homo sapiens*. *Journal of Cosmology*, **24**, 12019–12031 (http://journalofcosmology.com/JOC24/Tok_Pans_Paper_2-3.pdf).

EPILOGUE

As I approach the end of this narrative I ask myself many questions. Have I achieved what I had set out to achieve at the start of my journey? Have I felt the exhileration of discovery, the glory of new knowledge? Have I experienced the joys of friendship and love on the way to achieving my goals? To all these questions I must answer in the affirmative. I was fortunate to have parents who cared for me and whom I loved. My father was an inspiration in my career from a very early age. My mother devoted her life to my care and nurture throughout childhood. I was fortunate in the teachers I had at school and at University who inspired and encouraged me.

In later life I was lucky in the extreme in my wife Priya — her devotion, companionship and love and unstinting support for me at all times. On a fateful day in February 1966 when Priya and I walked into the sunset along a beach in Ceylon dreaming of our life together and of a limitless future, little did we know then where our road would take us. We conjured in our imagination visions of a life on this enchanted island, but that of course was not to be. Our journey has taken us far and wide, both on this planet — and, in imagination at least, to the most distant places in the cosmos. But every moment of life was savoured and enjoyed, not regretted.

My childhood ambition to understand the graceful spectacle of the Milky Way, that greeted me night after night from the veranda of my home in Ceylon has mostly been realised. I enjoyed the friendship and cooperation of colleagues, and in particular I formed a friendship and close working relationship with my mentor Fred Hoyle, whom I consider to be amongst the greatest astronomers and thinkers of all time. I have made my own humble contributions to astronomy in unravelling the organic nature of cosmic dust — those myriads of dust particles that can be seen as dark obscuring clouds against the background of stars in the Milky Way.

In my adolescent years I penned this simple Haiku-style verse, with an amazingly prophetic significance to my later life and career:

> "Amongst a myriad stars
> I stand alone....
> And wonder
> How much life and love
> Is there tonight...."

That sense of wonder has turned into realisation and understanding. My studies from the 1980's onwards that attempted to connect cosmic dust with life was considered controversial for a long time, but now many aspects of this work are coming to the fore and being accepted as correct. Early ideas that were once branded as unorthodox are moving imperceptibly into the domain of the orthodox. I am confident that a full and unequivocal recognition of our cosmic origins is round the corner. New vistas of knowledge are beginning to open before us, which would be the privilege of future generations — our grandchildren — to explore.

Printed in the United States
By Bookmasters